Interpretation of Algebraic Inequalities

Interpretation of Algebraic Inequalities

Practical Engineering Optimisation and Generating New Knowledge

Michael Todinov

CRC Press is an imprint of the
Taylor & Francis Group, an **informa** business

First edition published 2022
by CRC Press
6000 Broken Sound Parkway NW, Suite 300, Boca Raton, FL 33487-2742

and by CRC Press
2 Park Square, Milton Park, Abingdon, Oxon, OX14 4RN

ISBN: 9781032059174 (hbk)
ISBN: 9781032059198 (pbk)
ISBN: 9781003199830 (ebk)

DOI: 10.1201/9781003199830

Typeset in Times
by codeMantra

To Prolet and Marin

Contents

Preface

Algebraic inequalities are an ideal tool for handling deep uncertainty associated with components, properties and control parameters. A formidable advantage of algebraic inequalities is that they *do not require knowledge related to the values or distributions of the variables entering the inequalities.* This advantage permits ranking of systems and processes with unknown performance characteristics of their components.

A treatment related to the applications of algebraic inequalities for risk and uncertainty reduction was presented in a book, recently published by the author, focused exclusively on the *forward approach* to generating new knowledge by using algebraic inequalities. This approach starts with the competing physical systems/processes and compares their performance by deriving and proving algebraic inequalities. The forward approach was demonstrated through many application examples from various unrelated domains. However, the potential of the forward approach for generating new knowledge is somewhat limited because it is confined only to specific performance characteristics or specific systems.

The focus of this book is on the *inverse approach*, which was developed to expand the capabilities of the forward approach. Instead of comparing systems and processes and deriving and proving algebraic inequalities demonstrating the superiority of one of the solutions, the inverse approach operates in the opposite direction. The inverse approach starts with a correct algebraic inequality and progresses towards deriving new knowledge about a real system or process, which is subsequently used for enhancing its performance.

The inverse approach is based on the principle of non-contradiction: *if the variables and the different terms of a correct algebraic inequality can be interpreted as parts of a system or process, in the physical world, the system or process exhibits properties or behaviours that are consistent with the prediction of the algebraic inequality.*

This book demonstrates that powerful quantitative knowledge about systems and processes, locked in abstract inequalities, can be *released through their meaningful interpretation.* Furthermore, depending on the specific interpretation, knowledge, applicable to different systems from different domains, *can be released from the same starting inequality.* In this respect, the inverse approach *does not require or imply any forward analysis* of pre-existing systems or processes. The new properties discovered about real systems or processes are solely a result of meaningful interpretation of the variables and the different parts of the inequalities.

The book is firmly rooted in the practical applications of algebraic inequalities. In engineering design, design inequalities have been widely used to express design constraints guaranteeing that the design will perform its required functions. As will be shown in this book, the application potential of algebraic inequalities in engineering is far-reaching and certainly not restricted to specifying design constraints. Interpretation with the purpose of deriving new knowledge, subsequently used for system/process optimisation, is a new dimension in the application of algebraic inequalities. Thus, in Chapter 3, meaningful interpretation of algebraic inequalities was used to (i) construct a parallel–series system with superior reliability, (ii) select a complex system with superior reliability in the absence of knowledge related to the reliabilities of the separate components and (iii) rank alternative mechanical assemblies in terms of equivalent stiffness in the absence of data related to the stiffness of the individual components.

Practical applications of interpretation of algebraic inequalities are featured in the domains of reliability engineering, risk management, mechanical engineering, electrical engineering, economics, operational research and project management. The meaningful interpretation of algebraic inequalities, however, is not restricted to these areas only. Chapters 4 and 5 demonstrate an important class of algebraic inequalities based on sub- and super-additive functions which *can be used for optimising systems and processes in any area of science and technology, as long as the variables and the separate terms in the inequalities are additive quantities.*

By interpreting inequalities based on sub-additive functions in Chapters 4 and 5, new knowledge is derived and subsequently used to (i) produce light-weight designs, (ii) maximise the total power output from a voltage source, (iii) maximise the energy stored in a capacitor, (iv) maximise the accumulated elastic strain energy during tension and bending, (v) maximise the accumulated kinetic energy during inelastic impact and (vi) maximise the profit from investment. The same classes of algebraic inequalities have also been used to minimise the drag force during motion in viscous fluid and the formation of undesirable brittle phase during phase transformation.

In addition, the meaningful interpretation of a new algebraic inequality led to a method for maximising the mass of deposited substance during electrolysis.

Chapter 6 presents interpretation of algebraic inequalities for (i) maximising the probability of selecting high-reliability items from suppliers with unknown proportions of high-reliability items, (ii) optimal assignment of devices to tasks to maximise the probability of successful accomplishment of a mission and (iii) assessing the likelihood of unsatisfied demand from users placing random demands on a time interval.

Chapter 7 features interpretation of algebraic inequalities for (i) ranking the magnitudes of sequential random events, (ii) determining the lower bound of the probability of reliable assembly, (iii) determining tight lower and upper bound for the fraction of faulty components in a pooled batch and (iv) avoiding overestimation of profit.

Finally, Chapter 8 introduces interpretation of algebraic inequalities as potential energy to determine the equilibrium state of systems and sum of distances.

This book is the first publication introducing generation of new knowledge by meaningful interpretation of algebraic inequalities. It can be a primary source for a course on algebraic inequalities and their applications and a primary reading for researchers wishing to make a fast and powerful impact by creating new knowledge about systems and processes through interpretation of algebraic inequalities. This book can be an excellent source of applications for courses and projects in mathematical modelling, applied mathematics, reliability engineering, mechanical engineering, electrical engineering, operational research and risk management.

Finally, by demonstrating the principle of non-contradiction on numerous examples, this book sheds some light on the deep connection between the physical reality and mathematics. The results presented in this book support the view that physical phenomena and processes seem to take (follow) paths that are consistent with the predictions of correct abstract algebraic inequalities whose variables and different parts correspond to the controlling variables and factors driving these phenomena and processes.

In conclusion, I thank the Taylor & Francis Group Mechanical Engineering Editor Nicola Sharpe, the production editor Robert Sims and the project manager Saranya Narayanan from codeMantra for their valuable help and cooperation. Thanks also go to my academic colleagues for the useful comments on various aspects of the presented results.

Finally, I acknowledge the immense help and support from my wife Prolet, during the preparation of this book.

Michael Todinov
Oxford, June, 2021

Author

Michael Todinov has a background in mechanical engineering, applied mathematics and computer science. After receiving a PhD and a higher doctorate DEng from the University of Birmingham, he built an international reputation working on reliability and risk, repairable flow networks, probabilistic models in engineering, probabilistic fatigue and fracture and applications of algebraic inequalities in science and technology. He is a professor of mechanical engineering at Oxford Brookes University, UK.

Author

Michael [illegible] has [illegible] in mechanical engineering [illegible] Science [illegible] Mechanical [illegible] from the University of Birmingham. He holds [illegible] international reputation [illegible] in reliability and [illegible] and fracture [illegible] Brookes University, UK.

1 Fundamental Approaches in Modelling Real Systems and Processes by Using Algebraic Inequalities. The Principle of Non-contradiction for Algebraic Inequalities

1.1 ALGEBRAIC INEQUALITIES AND THEIR GENERAL APPLICATIONS

Algebraic inequalities have been used extensively in mathematics and a number of useful non-trivial algebraic inequalities and their properties have been well documented (Bechenbach and Bellman, 1961; Cloud et al., 1998; Engel, 1998; Hardy et al., 1999; Pachpatte, 2005; Steele, 2004; Kazarinoff, 1961; Sedrakyan and Sedrakyan, 2010). A comprehensive overview on the use of inequalities in mathematics has been presented in Fink (2000).

For a long time, simple inequalities are being used to express error bounds in approximations and constraints in linear programming models. In reliability and risk research, inequalities have been used as a tool for characterisation of reliability functions (Ebeling, 1997; Xie and Lai, 1998; Makri and Psillakis, 1996; Hill et al., 2013; Berg and Kesten, 1985; Kundu and Ghosh, 2017; Dohmen, 2006) and for reducing uncertainty and risk (Todinov, 2020a).

Applications of inequalities have been considered in physics (Rastegin, 2012) and engineering (Cloud et al., 1998; Samuel and Weir, 1999).

In engineering design, design inequalities have been used widely to express design constraints guaranteeing that the design will perform its required function (Samuel and Weir, 1999). This book shows that the application potential of algebraic inequalities in engineering is far reaching and certainly not restricted to specifying design constraints.

DOI: 10.1201/9781003199830-1

The method of algebraic inequalities is a domain-independent reliability-improvement method which derives from the domain-independent nature of mathematics. Applications of algebraic inequalities can be demonstrated in such diverse domains as mechanical engineering, electrical engineering, optimisation, operational research, project management, economics, decision-making under uncertainty, manufacturing and quality control.

A formidable advantage of the method of algebraic inequalities introduced in Todinov (2020a) is that algebraic inequalities *do not require knowledge related to the values or distributions of the variables entering the inequalities.* This makes the method of algebraic inequalities ideal for comparing design alternatives and processes in the absence of knowledge related to values of properties and control parameters.

1.2 ALGEBRAIC INEQUALITIES AS A DOMAIN-INDEPENDENT METHOD FOR REDUCING UNCERTAINTY AND OPTIMISING THE PERFORMANCE OF SYSTEMS AND PROCESSES

The conventional approaches for handling uncertainty in reliability engineering, for example, rely heavily on probabilities (Ramakumar, 1993; Lewis, 1996; Ebeling, 1997; Hoyland and Rausand, 1994; Dhillon, 2017; Todinov, 2002a, 2006a, b). These approaches effectively deal with structured uncertainty and the required probabilities are usually assessed by using a data-driven approach or by using the Bayesian, subjective probability approach.

The major deficiency of the data-driven approach is that probabilities cannot always be meaningfully defined. To be capable of making reliable predictions, for example, the data-driven approach needs past failure rates. Models based on failure rate data collected for a particular environment (temperature, humidity, pressure, vibrations, corrosive agents, etc.), however, give poor predictions for the time to failure in a different environment.

The deficiencies of the data-driven approach in reliability engineering cannot be rectified by using the Bayesian approach which is not so critically dependent on the availability of past failure data since it uses subjective, knowledge-based probabilities expressing the degree of belief about the outcome of a random event (Winkler, 1996). The subjective probabilities are subsequently updated as new experimental evidence becomes available (Ang and Tang, 2007). The Bayesian approach, however, depends on a selected probability model that may not be relevant to the modelled phenomenon/process (Aven, 2017). In addition, the assigned subjective probabilities depend on the available knowledge and vary significantly among the assessors. In this respect, weak background knowledge underlying the assigned subjective probabilities often results in poor predictions. Although the info-gap theory (Haim, 2005) deals with unstructured uncertainty by not making probability distribution assumptions, it still requires assumptions to be made about designer's best estimate.

In using algebraic inequalities to handle unstructured uncertainty, there is no need to assign frequentist probabilities, subjective probabilities or any particular probabilistic models. In addition, the use of algebraic inequalities requires no assumptions about estimates of parameters and property values. Algebraic inequalities *do not require knowledge related to the distributions of the variables entering the inequalities or any other assumptions*, and this makes the method of algebraic inequalities ideal for handling deep unstructured uncertainty associated with components, properties and values of control parameters.

Although the probabilities remain unknown, the algebraic inequalities can still establish the intrinsic superiority of one of the competing options. In this respect, algebraic inequalities avoid a major difficulty in the conventional models for handling uncertainty – lack of meaningful specification of frequentist probabilities or weak knowledge behind the assigned subjective probabilities and probabilistic models. Furthermore, the method of algebraic inequalities is also capable of discovering new properties related to systems and processes.

While reliability and risk assessment (Henley and Kumamoto, 1981; Kaplan and Garrick, 1981; Vose, 2000; Aven, 2003; Todinov, 2007) are truly domain-independent areas, this cannot be stated about the equally important areas of reliability improvement and risk reduction. For decades, the reliability and risk science failed to appreciate and emphasise that reliability improvement, risk and uncertainty reduction are underpinned by general principles that work in many unrelated domains.

As a consequence, *methods for measuring and assessing reliability, risk and uncertainty were developed, not domain-independent methods for improving reliability, reducing risk and uncertainty which could provide a direct input in the design process.* Indeed, in standard textbooks on mechanical engineering and design of machine components (French, 1999; Thompson, 1999; Collins, 2003; Norton, 2006; Pahl et al., 2007; Childs, 2014; Budynas and Nisbett, 2015; Mott et al., 2018; Gullo and Dixon, 2018), for example, there is practically no discussion of generic (domain-independent) methods for reliability improvement and risk and uncertainty reduction.

The problem is that the current approach to reliability improvement and risk reduction almost solely relies on knowledge from a specific domain and is conducted exclusively by experts in that domain. This creates the incorrect perception that effective reliability improvement and risk reduction can be delivered only by using methods offered by the specific domain, without resorting to general risk reduction methods and principles.

This incorrect perception resulted in ineffective reliability improvement and risk reduction, the loss of valuable opportunities for reducing risk and repeated 'reinvention of the wheel'. Current technology changes so fast that the domain-specific knowledge related to reliability improvement and risk reduction is outdated almost as soon as it is generated. In contrast, the domain-independent methods for reliability improvement, risk and uncertainty reduction are higher-order generic methods that permit application in new, constantly changing situations and circumstances.

A central theme in the domain-independent approach for reliability improvement and risk reduction introduced in Todinov (2019a, b) is the concept that

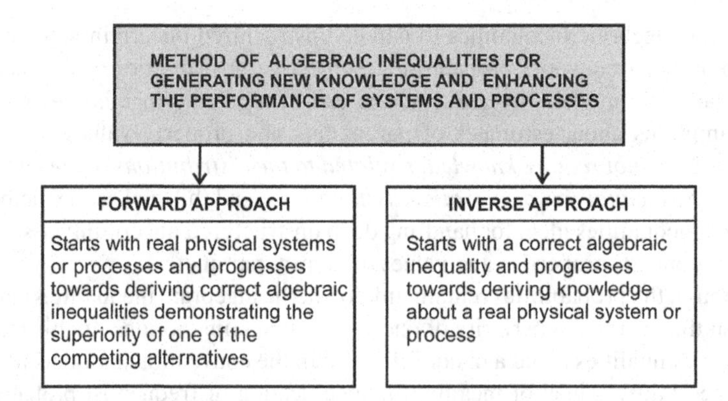

FIGURE 1.1 Two basic approaches of the method of algebraic inequalities for generating new knowledge and enhancing system and process performance.

risk reduction is underlined by common domain-independent principles which, if combined with knowledge from a specific domain, (i) provide key input to the design process by improving the reliability of the designed product, (ii) avoid loss of opportunities for improving reliability and reducing risk and (iii) provide effective risk and uncertainty reduction across unrelated domains of human activity.

As part of the domain-independent approach for reliability improvement and risk reduction, a number of new domain-independent methods have been introduced in Todinov (2019a, b): *separation, segmentation, self-reinforcement, introducing deliberate weaknesses, inversion, reducing the rate of damage accumulation, comparative reliability models, permutation and substitution.* The method of algebraic inequalities is another powerful domain-independent method for generating new knowledge about systems and processes that can be used for enhancing their performance.

There are two major approaches of the method of algebraic inequalities: (i) *forward approach*, which starts with real systems and processes and progresses towards deriving correct algebraic inequalities demonstrating the superiority of one of the competing alternatives and (ii) *inverse approach*, which consists of interpreting a correct abstract inequality and inferring from it unknown properties related to a real physical system or process (Figure 1.1).

1.3 FORWARD APPROACH TO MODELLING AND OPTIMISATION OF REAL SYSTEMS AND PROCESSES BY USING ALGEBRAIC INEQUALITIES

The forward approach includes several basic steps (Figure 1.2): (i) analysis of the system or process, (ii) conjecturing inequalities ranking the competing alternatives and (iii) testing and proving rigorously the conjectured inequalities and

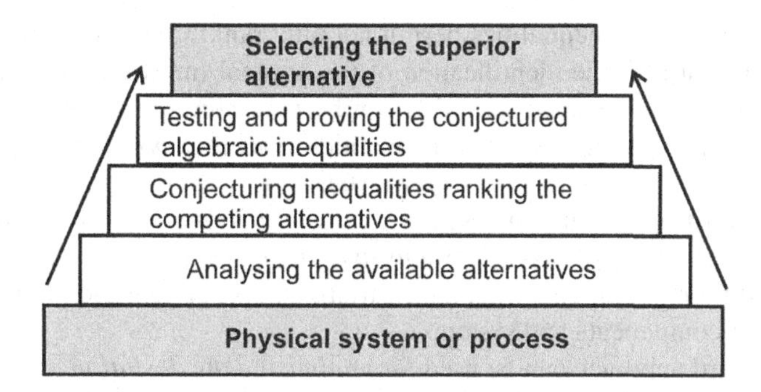

FIGURE 1.2 Forward approach of the method of algebraic inequalities.

selecting the superior competing alternative. The forward approach for generating knowledge and improving systems and process performance has already been explored in Todinov (2019c, d, 2020a, b, c, d).

By following this approach, algebraic inequalities, for example, can be used to *rank systems with unknown reliabilities of their components.*

The generic strategy in comparing reliabilities of competing systems starts with building the functional diagrams of the systems, creating reliability networks based on the functional diagrams, deriving expressions for the system reliability of the competing alternatives, conjecturing inequalities related to the competing alternatives, testing the conjectured inequalities by using Monte Carlo simulation and developing rigorous proofs by using some combination of analytical techniques (Figure 1.3).

As a result, the reliabilities of two systems can be compared in the absence of knowledge about the reliabilities of the separate components or in the presence of partial knowledge only. Partial knowledge is present if, for example, it is known that a particular component is older (less reliable) than another component of the

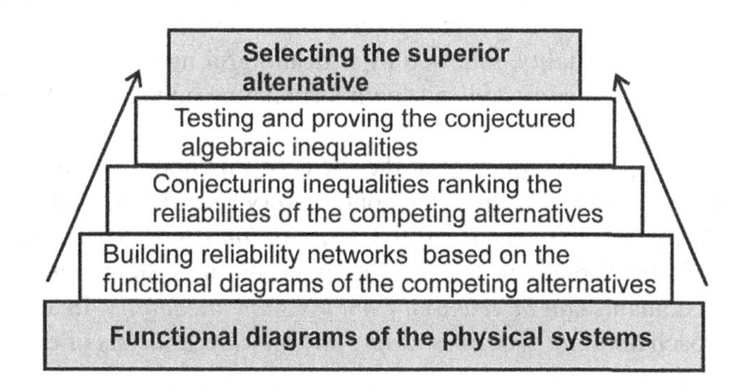

FIGURE 1.3 Forward approach used for ranking the reliabilities of systems with unknown reliabilities of their components.

same type. Algebraic inequalities permit not only ranking two systems in terms of reliability but also the identification of the optimal (most reliable) system. If all available topologies are listed, this can be done through series of algebraic inequalities. Such an application of inequalities to achieve system reliability optimisation has been discussed in Todinov (2021).

The forward approach can also be used *for determining upper and lower bounds for critical properties of systems and processes*. In this way, inequalities can be used for reducing the worst-case variation of properties and improving the reliability of components and systems.

The forward approach can be used *for minimising the deviation of reliability-critical parameters from their required values, for maximising the system reliability* and *for minimising the risk of a faulty assembly* and these applications have also been demonstrated in Todinov (2020a).

1.4 INVERSE APPROACH TO MODELLING AND GENERATING NEW KNOWLEDGE BY INTERPRETATION OF INEQUALITIES

The treatment related to the applications of algebraic inequalities (Todinov, 2020a) for reducing uncertainty and risk was based exclusively on the forward approach, outlined in Section 1.3. The forward approach is very powerful, but its potential for generating new knowledge is somewhat limited. The comparison of competing systems/processes does not always yield algebraic inequalities that hold for all values of the controlling variables. In some cases, the conjectured inequality holds for some set of values for the controlling variables, but for another set of values, the conjectured inequality does not hold. Because of this, in some cases, no intrinsically more reliable system or process emerges as a result of the forward approach. Furthermore, the potential for generating new knowledge of the forward approach is confined to the specific systems and processes that are compared.

These limitations can be overcome by the inverse approach, which has been developed in Todinov (2020b; 2021). The inverse approach starts with a correct algebraic inequality, progresses through creating, relevant meaning for the variables entering the inequality, followed by a meaningful interpretation of the different parts of the inequality and ends with formulating undiscovered properties/knowledge about the system or process (Figure 1.4).

The inverse approach, which is at the focus of this book, is founded on the observation that abstract inequalities contain very useful knowledge that can be *released through their meaningful interpretation*. Furthermore, depending on the specific interpretation, knowledge applicable to systems and processes from diverse domains *can be released from the same inequality*. In addition, the inverse approach *does not require or imply any forward analysis* of existing systems or processes.

An important step of the inverse approach is creating, relevant meaning for the variables entering the algebraic inequality, followed by a meaningful interpretation of the different parts of the inequality and releasing new knowledge.

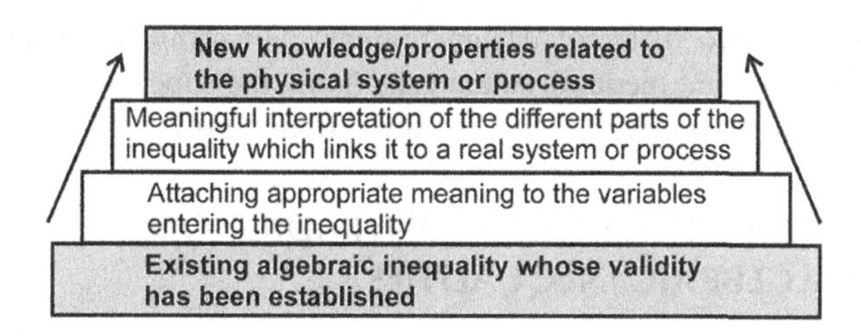

FIGURE 1.4 Inverse approach to deriving new knowledge by using algebraic inequalities.

The inverse approach effectively links existing correct abstract algebraic inequalities with real physical systems or processes and opens opportunities for enhancing their performance through discovering new fundamental properties.

The initial step of the inverse approach is testing an existing inequality (e.g. by using Monte Carlo simulation) (Figure 1.4). If no combinations of input data contradicting the inequality have been found during the testing stage, the conjectured inequality is plausible and a rigorous proof is attempted. Next, relevant meaning for the variables entering the inequality is created, followed by a meaningful interpretation of its parts (Figure 1.4).

A trivial algebraic inequality can certainly be interpreted meaningfully, but the knowledge extracted from this interpretation is obvious and can also be easily reached intuitively. In contrast, the interpretation of a non-trivial algebraic inequality, for example, the inequality $x^2 + y^2 + z^2 \geq xy + yz + zx$, where x, y and z are any real numbers, leads to deep insights that are not at all obvious and cannot be reached intuitively. In fact, as Section 6.1 demonstrates, in some cases, the knowledge obtained from interpreting non-trivial inequalities can be rather counter-intuitive.

The key idea of this book is that non-trivial algebraic inequalities can be interpreted meaningfully and the new knowledge extracted from the interpretation can be used for optimising systems and processes in diverse areas of science and technology. The knowledge extracted from the interpretation of non-trivial inequalities is non-trivial and cannot be reached intuitively.

Interpretation of abstract inequalities led to the discovery of a number of overlooked useful properties in such mature fields like mechanical engineering, electrical engineering, reliability engineering and risk management.

The inverse approach always leads to new results, as long as a meaningful interpretation of the variables and the different parts of the inequalities can be done. This is because the inverse approach is rooted in the principle of non-contradiction: *if the variables and the different terms of a correct algebraic inequality can be interpreted as parts of a system or process, in the real world, the system or process exhibits properties or behaviour that are consistent with the prediction of the algebraic inequality.* In short, the realisation of the process/experiment yields results that do not contradict the algebraic inequality.

A central idea in optimising designs by using algebraic inequalities is to interprete the left- and right-hand side of a correct algebraic inequality as a particular output related to two different design options, delivering the same required function. The algebraic inequality then establishes the superiority of one of the compared design options with respect to the chosen output.

1.5 THE PRINCIPLE OF NON-CONTRADICTION FOR ALGEBRAIC INEQUALITIES

The link between physical reality and algebraic inequalities can be demonstrated with the next simple example. Consider the common abstract inequality

$$a^2 + b^2 \geq 2ab \tag{1.1}$$

which is true for any real numbers a, b because the inequality can be obtained from the obvious inequality $(a - b)^2 \geq 0$. Adding $2ab$ to both sides of inequality (1.1) will not change its direction and the result is the inequality:

$$(a + b)^2 \geq 4ab \tag{1.2}$$

Consider the positive quantities a, b $(a > 0, b > 0)$. Dividing both sides of inequality (1.2) by the positive value $(a + b)$ does not alter the direction of the inequality:

$$a + b \geq 4\frac{ab}{a + b} \tag{1.3}$$

Dividing the numerator and denominator of the right-hand side of equation (1.3) by ab gives

$$a + b \geq 4\frac{1}{1/a + 1/b} \tag{1.4}$$

The left- and right-hand side of inequality (1.4) can now be interpreted by noticing that if the variables a and b stand for the resistances of two elements, the left-hand side of inequality (1.4) can be interpreted as the equivalent resistance of the elements connected in series (Figure 1.5a). The right-hand side of inequality (1.4) can be interpreted as the equivalent resistance of the same elements connected in parallel, multiplied by 4 (Figure 1.5b).

Inequality (1.4) predicts that *the equivalent resistance of two elements connected in series is at least four times greater than the equivalent resistance of the same elements connected in parallel, irrespective of the individual resistances of the elements.*

If physical measurements of the equivalent resistances of the arrangements in Figure 1.5a and b are conducted, they will only confirm the prediction from the algebraic inequality: that for any combination of values a and b for the resistances of the two elements, the equivalent resistance in series is always at least four times greater than the equivalent resistance of the same elements in parallel.

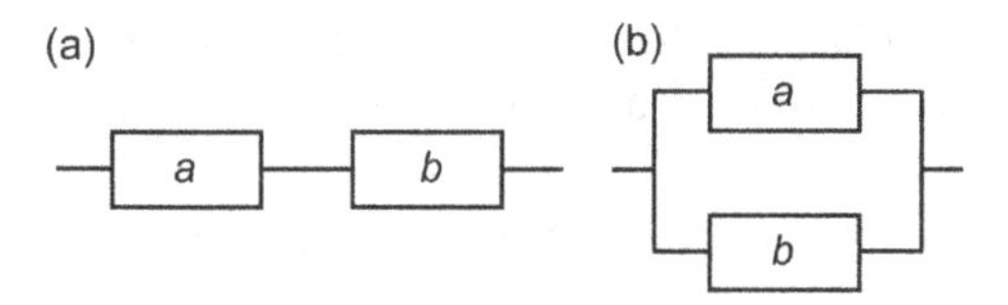

FIGURE 1.5 Parts of electrical circuit demonstrating the two-way connection between algebraic inequalities and physical reality; (a) resistors connected in series and (b) resistors connected in parallel.

The examples in this book demonstrate a similar perfect agreement of predictions from abstract inequalities and properties and behaviour of physical phenomena and processes. In rationalising the agreement of experimental results and predictions from abstract inequalities, a natural question arises as to why the real world complies its response with what an abstract inequality predicts.

The debate about the unreasonable efficiency of mathematics in the natural sciences started with the famous paper of Wigner (1960). Recent decades have seen a continuation of the discourse (Penrose, 1989; Livio, 2009; Vedral, 2010; Tegmark, 2014) related to the old question dating back to Pythagoras and Plato about whether mathematics is a human invention or expresses truths with independent existence that are discovered. The positions in this vigorous debate range from 'mathematics is a pure human invention' – a set of rules for manipulating a set of symbols to the view: 'reality is mathematics' (Tegmark, 2014).

The results presented in this book support the view that physical phenomena/processes take (follow) paths that are consistent with the predictions of correct abstract algebraic inequalities whose variables and different parts correspond to key variables and parameters controlling these phenomena/processes. This is the essence of the principle of non-contradiction which underlies deriving knowledge from a meaningful interpretation of algebraic inequalities.

It seems that mathematics is interwoven in the fabric of the physical world and the course of physical phenomena and processes is inherently consistent with the relevant algebraic inequalities.

The law of non-contradiction between the predictions of relevant algebraic inequalities and the development of real processes not only infers the existence of particular properties but also forbids the existence of properties that are not in agreement with the relevant algebraic inequalities. To illustrate this point, consider two objects and a physical process where the probability of appearance of a particular feature X in a selected object is $P(X) = p$. Consequently, the probability of non-appearance of the feature X in a selected object is $P(\overline{X}) = 1 - p$. In addition, the probability of appearance/non-appearance of the feature in one of the objects does not depend on the appearance/non-appearance of the feature in the other object. This means that the appearance/non-appearance of the feature in the individual objects are statistically independent events.

In this case, the probability that the two objects will either both have the feature X or both will not have the feature X is

$$P(X \cap X \cup \bar{X} \cap \bar{X}) = p^2 + (1-p)^2 \tag{1.5}$$

The probability that one of the objects will have the feature and the other object will not is

$$P(X \cap \bar{X} \cup \bar{X} \cap X) = p(1-p) + (1-p)p = 2p(1-p) \tag{1.6}$$

Suppose that it has been conjectured that the appearance/non-appearance of the feature in the individual objects are statistically independent events. In this case, the inequality

$$p^2 + (1-p)^2 \geq 2p(1-p) \tag{1.7}$$

follows directly from the obvious inequality $[p-(1-p)]^2 \geq 0$, irrespective of the probability p. If the appearance/non-appearance of the feature in each of the objects are statistically independent events, inequality (1.7) must hold.

Let a particular experiment yield empirically that the probability of both objects having the feature or both objects not having the feature is smaller than the probability that one object will have the feature and the other object will not: $P(X \cap X \cup \bar{X} \cap \bar{X}) < P(X \cap \bar{X} \cup \bar{X} \cap X)$. In this case, inequality (1.7) does not hold for that experiment, which means that the appearance/non-appearance of the feature in the individual objects cannot be statistically independent events.

1.6 KEY STEPS IN THE INTERPRETATION OF ALGEBRAIC INEQUALITIES

The key steps of the interpretation of algebraic inequalities are (i) identifying classes of abstract inequalities which permit interpretation that can be linked easily with a real system or process and (ii) identifying transformations that make the inequality interpretable.

Very important candidates for meaningful interpretation are the wide class of inequalities based on *super-* and *sub-additive* functions. If a particular sub- or super-additive function measures the effect of a particular additive factor, the corresponding inequality has the potential to significantly increase the effect of the factor by segmenting it or by aggregating it.

Additive quantities can be found in all areas of science and engineering. Additive quantities change with changing the size of their supporting objects/systems. Examples of additive quantities are mass, weight, amount of substance, number of particles, volume, distance, energy (kinetic energy, gravitational energy, electric energy, elastic energy, surface energy, internal energy), work, power, heat, force, momentum, electric charge, electric current, heat capacity, electric capacity, resistance (when the elements are in series), enthalpy and fluid flow.

The meaningful interpretation of algebraic inequalities, however, is not restricted to inequalities that only provide segmentation or segregation of additive controlling factors. For example, the meaningful interpretation of the well-known arithmetic mean–harmonic mean algebraic inequality helped reveal important properties of parallel and series arrangements of mechanical, electrical and thermal components.

The meaningful interpretation of a new class of algebraic inequalities proposed in this book helped improve the performance of parallel–series systems. In addition, the meaningful interpretation of the inequality of negatively correlated events helped reveal the system with superior intrinsic reliability.

Knowledge released from meaningful interpretation of algebraic inequalities is relevant to various application domains and helped find overlooked properties and opportunities for enhancing systems performance in mature fields such as reliability engineering, risk management, mechanical engineering, electrical engineering, management and economics.

The potential application of algebraic inequalities is very broad, but the areas of risk and reliability are a particularly important application domains. Thus, the generated new knowledge in the area of reliability and risk through interpretation of algebraic inequalities provides the opportunity to construct systems with superior intrinsic reliability in the absence of knowledge related to the failure rates of the components building the system.

In the area of mechanical engineering, generating new knowledge through interpretation of algebraic inequalities provided the opportunity to develop lightweight designs and maximise the capacity for elastic strain energy accumulation of various mechanical assemblies. In addition, new knowledge generated through interpretation of algebraic inequalities provided the opportunity to optimise the performance of mechanical systems.

In the area of electrical engineering, new knowledge generated from interpreting algebraic inequalities provided the possibility to maximise the power output in electrical circuits and the energy stored in capacitors.

In the area of risk management, knowledge generated through interpretation of algebraic inequalities provided the opportunity to maximise the profits from an investment, avoid risk underestimation and maximise the likelihood of correct ranking the magnitudes of sequential uncertain events.

Knowledge generated from the interpretation of new algebraic inequalities proposed in this book can also be used *for increasing significantly the effect of additive quantities (factors).*

The new results derived by interpreting algebraic inequalities have not been stated in any publication in the mature fields of mechanical engineering, reliability engineering, risk management and electrical engineering which demonstrates that the lack of knowledge of the method of algebraic inequalities made these important results invisible to domain experts.

2 Basic Algebraic Inequalities

2.1 BASIC ALGEBRAIC INEQUALITIES USED FOR PROVING OTHER INEQUALITIES

2.1.1 Basic Properties of Algebraic Inequalities and Techniques for Proving Algebraic Inequalities

Inequalities are statements about expressions or numbers which involve the symbols '<' (less than), '>' (greater than), '≤' (less than or equal to) or '≥' (greater than or equal to).

The basic rules related to handling algebraic inequalities can be summarised as follows:

a. For any real numbers a and b, exactly one of the following holds:

$$a < b,\ a = b,\ a > b.$$

b. If $a > b$ and $b > c$, then $a > c$
c. If $a > b$, adding the same number c to both sides of the inequality does not alter its direction: $a + c > b + c$
d. Multiplying both sides of an inequality by (-1) reverses the direction of the inequality:

$$\text{if } a > b, \text{ then } -a < -b; \text{ if } a < b, \text{ then } -a > -b$$

e. If $a > 0$ and $b > 0$, then $ab > 0$

By using the basic rules, the next basic properties can be established (see Todinov, 2020a for more details and proofs)

i. For any real number x, $x^2 \geq 0$. The equality holds if and only if $x = 0$.
ii. If $x > y, t > 0$, then $xt > yt$ and $x/t > y/t$.
   From this property, it follows that if $0 < x < 1$, then $x^2 < x$.
iii. If $x > y > 0, u > v > 0$, then $xu > yv$ and $x/v > y/u$.
   From this property, it follows that if $a > b > 0$, then $a^2 > b^2$.
iv. If $x > 0, y > 0, x \neq y$ and $x^2 > y^2$, then $x > y$.

DOI: 10.1201/9781003199830-2

v. If a strictly increasing function is applied to both sides of an inequality, the inequality will still hold. Applying a strictly decreasing function to both sides of an inequality reverses the direction of the inequality.
   Thus, from $x > y > 0$, it follows that $\ln x > \ln y$, $x^n > y^n$, where $n > 0$.
   From $x > y > 0$, it follows that $x^{-n} < y^{-n}$, where $n > 0$.

A comprehensive treatment of various methods for proving algebraic inequalities has been presented in Todinov (2020a) which covers the following techniques:

- Direct algebraic manipulation and analysis
- Using the properties of convex/concave functions
- Segmentation through basic inequalities
- Transforming algebraic inequalities to known basic inequalities
- Strengthening the inequalities
- Conditioning on mutually exclusive cases
- Exploiting the symmetry of the variables entering the inequality
- Using mathematical induction
- Using the properties of sub- and super-additive functions
- Using substitution
- Exploiting homogeneity
- Using derivatives

Most of these techniques are used for proving inequalities in this book.

2.1.2 Cauchy–Schwarz Inequality

An important basic algebraic inequality with numerous applications is the Cauchy–Schwarz inequality (Cauchy, 1821; Steele, 2004) which states that for the sequences of real numbers $a_1, a_2, \ldots, a_n$ and $b_1, b_2, \ldots, b_n$, the following inequality holds:

$$\left(a_1b_1 + a_2b_2 + \cdots + a_nb_n\right)^2 \le \left(a_1^2 + a_2^2 + \cdots + a_n^2\right)\left(b_1^2 + b_2^2 + \cdots + b_n^2\right) \quad (2.1)$$

Equality holds if and only if, for any $i \ne j$, $a_ib_j = a_jb_i$ are fulfilled.

The Cauchy–Schwarz inequality is a very powerful inequality and many algebraic inequalities can be proved by reducing them to the Cauchy–Schwarz inequality through appropriate substitutions.

The Cauchy–Schwarz inequality (2.1) can be proved by using direct algebraic manipulation and analysis based on the properties of the quadratic trinomial. Consider the expression

$$y = (a_1t + b_1)^2 + (a_2t + b_2)^2 + \cdots + (a_nt + b_n)^2 \quad (2.2)$$

For any sequences of real numbers $a_1, a_2, \ldots, a_n$ and $b_1, b_2, \ldots, b_n$, y is non-negative ($y \geq 0$). Expanding the right-hand side of equation (2.2) and collecting the coefficients in front of t^2 and t give the quadratic trinomial

$$y = \left(a_1^2 + a_2^2 + \cdots + a_n^2\right)t^2 + 2(a_1b_1 + a_2b_2 + \cdots + a_nb_n)t + \cdots + b_1^2 + b_2^2 + \cdots + b_n^2 \tag{2.3}$$

with respect to t. In Equation 2.3, y is non-negative ($y \geq 0$) only if $D \leq 0$, where D is the discriminant of the quadratic trinomial. Therefore, the condition

$$\left(a_1b_1 + a_2b_2 + \cdots + a_nb_n\right)^2 - \left(a_1^2 + a_2^2 + \cdots + a_n^2\right)\left(b_1^2 + b_2^2 + \cdots + b_n^2\right) \leq 0 \tag{2.4}$$

must hold for a non-negative y. Condition (2.4), however, is identical to the Cauchy–Schwarz inequality (2.1), which completes the proof.

2.1.3 Convex and Concave Functions: Jensen Inequality

A function $f(x)$ with a domain $[a, b]$ is said to be convex (Figure 2.1a), if for all values x_1 and x_2 in its domain $\left(x_1, x_2 \in [a, b]\right)$, the next inequality holds

$$f(wx_1 + (1-w)x_2) \leq wf(x_1) + (1-w)f(x_2) \tag{2.5}$$

where $0 \leq w \leq 1$.

If $f(x)$ is twice differentiable for all $x \in (a, b)$, and if the second derivative is non-negative ($f''(x) \geq 0$), the function $f(x)$ is convex on (a, b). For any convex function $f(x)$, the Jensen's inequality (Steele, 2004) states that

$$f(w_1x_1 + w_2x_2 + \cdots + w_nx_n) \leq w_1f(x_1) + w_2f(x_2) + \cdots + w_nf(x_n) \tag{2.6}$$

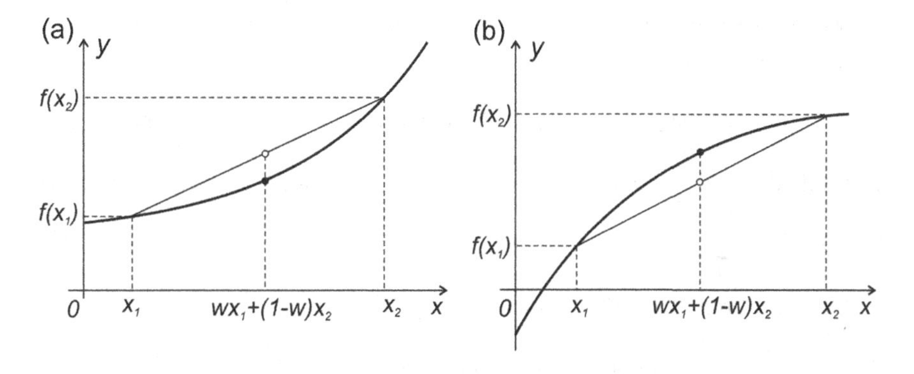

FIGURE 2.1 (a) A convex and (b) a concave function of a single argument.

where w_i $(i = 1,...,n)$ are numbers (weights) that satisfy $0 \le w_i \le 1$ and $w_1 + w_2 + \cdots + w_n = 1$.

If the function $f(x)$ is strictly convex (second derivative is positive), equality in inequality (2.6) is attained only for $x_1 = x_2 = ... = x_n$.

A function $f(x)$ with a domain $[a, b]$ is said to be concave (Figure 2.1b), if for all values x_1 and x_2 in its domain $(x_1, x_2 \in [a,b])$, the next inequality holds

$$f(wx_1 + (1-w)x_2) \ge wf(x_1) + (1-w)f(x_2) \tag{2.7}$$

where $0 \le w \le 1$.

For any concave function $f(x)$, the Jensen's inequality states that

$$f(w_1x_1 + w_2x_2 + \cdots + w_nx_n) \ge w_1f(x_1) + w_2f(x_2) + \cdots + w_nf(x_n) \tag{2.8}$$

where w_i $(i = 1,\ldots,n)$ are numbers (weights) that satisfy $0 \le w_i \le 1$ and $w_1 + w_2 + \cdots + w_n = 1$.

If $f(x)$ is twice differentiable for all $x \in (a,b)$, and if the second derivative is not positive $(f''(x) \le 0)$, the function $f(x)$ is concave on (a,b). If the function $f(x)$ is strictly concave (second derivative is negative), equality in inequality (2.8) is attained only for $x_1 = x_2 = \ldots = x_n$.

Proving an inequality by using the properties of concave functions will be demonstrated by proving the inequality

$$\left(\frac{m_1x_1 + m_2x_2 + \cdots + m_nx_n}{M}\right)^M \ge x_1^{m_1} x_2^{m_2} \ldots x_n^{m_n} \tag{2.9}$$

where $m_1,\ldots,m_n$ and $x_1,\ldots,x_n$ are positive values, and $M = \sum_{i=1}^{n} m_i$.

Consider the function $\ln x$ which is concave for positive values $x > 0$ because its second derivative $d^2(lnx)/dx^2 = -1/x^2$ is negative. According to the Jensen's inequality, for concave functions,

$$\ln(w_1x_1 + w_2x_2 + \cdots + w_nx_n) \ge w_1 \ln x_1 + w_2 \ln x_2 + \cdots + w_n \ln x_n$$

is fulfilled, where $w_i = m_i / M$. Note that w_i are treated as weights because $0 < w_i < 1$ and $\sum_{i=1}^{n} w_i = 1$.

Using the properties of the logarithms,

$$\ln(w_1x_1 + w_2x_2 + \cdots + w_nx_n) \ge \ln(x_1^{w_1} x_2^{w_2} \ldots x_n^{w_n})$$

Because both sides of the last inequality are positive, and e^x is an increasing function, from property (v) in Section 2.1.1, it follows that

$$\exp\left[\ln(w_1x_1 + w_2x_2 + \cdots + w_nx_n)\right] \geq \exp\left[\ln(x_1^{w_1}x_2^{w_2}\ldots x_n^{w_n})\right]$$

or finally,

$$w_1x_1 + w_2x_2 + \cdots + w_nx_n \geq x_1^{w_1}x_2^{w_2}\ldots x_n^{w_n} \quad (2.10)$$

Rising both sides of inequality (2.10) into the positive power of M, according to property (v) in Section 2.1.1, gives inequality (2.9):

$$\left(\frac{m_1x_1 + m_2x_2 + \cdots + m_nx_n}{M}\right)^M \geq \left(x_1^{m_1/M}x_2^{m_2/M}\ldots x_n^{m_n/M}\right)^M = x_1^{m_1}x_2^{m_2}\ldots x_n^{m_n}$$

2.1.4 Root-Mean Square–Arithmetic Mean–Geometric Mean–Harmonic Mean (RMS–AM–GM–HM) Inequality

For a set of positive real numbers $x_1, x_2, \ldots, x_n$, the basic RMS–AM–GM–HM inequality states

$$\sqrt{\frac{x_1^2 + x_2^2 + \cdots + x_n^2}{n}} \geq \frac{x_1 + x_2 + \cdots + x_n}{n} \geq \sqrt[n]{x_1x_2\ldots x_n} \geq \frac{n}{1/x_1 + 1/x_2 + \cdots + 1/x_n} \quad (2.11)$$

with equality attained only if $x_1 = x_2 = \ldots = x_n$.

The number $\sqrt{\frac{x_1^2 + x_2^2 + \cdots + x_n^2}{n}}$ is known as *root-mean square*; $\frac{x_1 + x_2 + \cdots + x_n}{n}$ is the *arithmetic mean*; $\sqrt[n]{x_1x_2\ldots x_n}$ is the *geometric mean* and $\frac{n}{1/x_1 + 1/x_2 + \cdots + 1/x_n}$ is the *harmonic mean.*

For two non-negative numbers a, b, the inequality takes the form

$$\sqrt{\frac{a^2 + b^2}{2}} \geq \frac{a + b}{2} \geq \sqrt{ab} \geq \frac{2}{\frac{1}{a} + \frac{1}{b}} \quad (2.12)$$

Various techniques for proving the RMS–AM–GM–HM inequalities have been demonstrated in Todinov (2020a). The AM–GM inequality, for example, can be obtained as a special case of inequality (2.9) for $m_1 = m_2 = \ldots = m_n = 1,\ \ M = \sum_{i=1}^{n} m_i = n$. As a result,

$$\left(\frac{m_1x_1 + m_2x_2 + \cdots + m_nx_n}{M}\right)^M = \left(\frac{x_1 + x_2 + \cdots + x_n}{n}\right)^n \geq x_1^{m_1} x_2^{m_2} \ldots x_n^{m_n}$$
$$= x_1x_2 \ldots x_n \tag{2.13}$$

or finally,

$$\frac{x_1 + x_2 + \ldots + x_n}{n} \geq \sqrt[n]{x_1x_2 \ldots x_n} \tag{2.14}$$

which completes the proof of the AM–GM inequality.

Segmentation through the AM–GM inequality (2.14) is a powerful technique for proving algebraic inequalities. The basic idea of this technique is to segment (split) the original inequality into simpler inequalities by using the AM–GM inequality and to sum the resultant inequalities in order to assemble the original inequality. Here is an example of this technique demonstrated on the inequality

$$x^2 + y^2 + z^2 \geq xy + yz + zx \tag{2.15}$$

valid for any arbitrary real x, y, z. This inequality can be segmented into three inequalities by using the standard AM–GM inequality:

$$\frac{x^2 + y^2}{2} \geq xy; \quad \frac{y^2 + z^2}{2} \geq yz; \quad \frac{z^2 + x^2}{2} \geq zx$$

Adding these three inequalities gives the original inequality (2.15) and completes the proof.

2.1.5 Rearrangement Inequality

The rearrangement inequality is a powerful yet underused basic algebraic inequality that can be applied to prove other inequalities.

Consider the two sequences $a_1, a_2, \ldots, a_n$ and $b_1, b_2, \ldots, b_n$ of real numbers. It can be shown that:

a. The sum $S = a_1b_1 + a_2b_2 + \cdots + a_nb_n$ is maximal if the sequences are sorted in the same way: both monotonically decreasing: $a_1 \geq a_2 \geq, \ldots, \geq a_n$; $b_1 \geq b_2 \geq, \ldots, \geq b_n$ or both monotonically increasing: $a_1 \leq a_2 \leq, \ldots, \leq a_n$; $b_1 \leq b_2 \leq, \ldots, \leq b_n$.
b. The sum $S = a_1b_1 + a_2b_2 + \cdots + a_nb_n$ is minimal if the sequences are sorted in the opposite way: one monotonically increasing and the other monotonically decreasing.

A proof of the rearrangement inequality can be found in Todinov (2020a).

The rearrangement inequality is a basis for generating new powerful inequalities that can be used to produce bounds for the uncertainty in reliability-critical parameters. For two sequences $a_1, a_2, \ldots, a_n$ and $b_1, b_2, \ldots, b_n$ of real numbers, the notation

$$\begin{bmatrix} a_1 & a_2 & \ldots & a_n \\ b_1 & b_2 & \ldots & b_n \end{bmatrix} = a_1 b_1 + a_2 b_2 + \cdots + a_n b_n$$

is introduced. This is similar to the definition of a dot product of two vectors with components specified by the two rows of the matrix.

An important corollary of the rearrangement inequality is the following:

- Given a set of real numbers $a_1, a_2, \ldots, a_n$, for any permutation $a_{1s}, a_{2s}, \ldots, a_{ns}$ of these numbers, the following inequality holds:

$$a_1^2 + a_2^2 + \cdots + a_n^2 \geq a_1 a_{1s} + a_1 a_{2s} + \cdots + a_{ns} \tag{2.16}$$

Indeed, without loss of generality, it can be assumed that $a_1 \leq a_2 \leq, \ldots, \leq a_n$. Applying the rearrangement inequality to the sequences $(a_1, a_2, \ldots, a_n)$, $(a_1, a_2, \ldots, a_n)$ and to the sequences $(a_1, a_2, \ldots, a_n)$, $(a_{1s}, a_{2s}, \ldots, a_{ns})$ then yields

$$\begin{bmatrix} a_1 & a_2 & \ldots & a_n \\ a_1 & a_2 & \ldots & a_n \end{bmatrix} = a_1^2 + a_2^2 + \cdots + a_n^2 \geq \begin{bmatrix} a_1 & a_2 & \ldots & a_n \\ a_{1s} & a_{2s} & \ldots & a_{ns} \end{bmatrix}$$
$$= a_1 a_{1s} + a_2 a_{2s} + \cdots + a_n a_{ns}$$

because the first pair of sequences are similarly ordered and the second pair of sequences are not. This completes the proof of inequality (2.16).

2.1.6 Chebyshev's Sum Inequality

Another important basic algebraic inequality is the Chebyshev's sum inequality (Besenyei, 2018). It states that for the sequences of real numbers $a_1 \geq a_2 \geq, \ldots, \geq a_n$ and $b_1 \geq b_2 \geq, \ldots, \geq b_n$, the following inequality holds:

$$n(a_1 b_1 + a_2 b_2 + \cdots + a_n b_n) \geq (a_1 + a_2 + \cdots + a_n)(b_1 + b_2 + \cdots + b_n)$$

or

$$\frac{a_1 b_1 + \cdots + a_n b_n}{n} \geq \frac{a_1 + \cdots + a_n}{n} \cdot \frac{b_1 + \cdots + b_n}{n} \tag{2.17}$$

If $a_1 \geq, \ldots, \geq a_n$ and $b_1 \leq, \ldots, \leq b_n$ hold, the inequality is reversed:

$$\frac{a_1 b_1 + \cdots + a_n b_n}{n} \leq \frac{a_1 + \cdots + a_n}{n} \cdot \frac{b_1 + \cdots + b_n}{n} \tag{2.18}$$

Equality is attained if $a_1 = a_2 = \ldots = a_n$ or $b_1 = b_2 = \ldots = b_n$ holds.

For the sequences of real numbers $a_1 \geq a_2 \geq, \ldots, \geq a_n$ and $b_1 \geq b_2 \geq, \ldots, \geq b_n$, Chebyshev's sum inequality (2.17) can be proved by using the rearrangement inequality discussed in the previous section. According to the rearrangement inequality, the following inequalities are true:

$$\sum_{i=1}^{n} a_i b_i = a_1 b_1 + a_2 b_2 + a_3 b_3 + \cdots + a_n b_n$$

$$\sum_{i=1}^{n} a_i b_i \geq a_1 b_2 + a_2 b_3 + a_3 b_4 + \cdots + a_n b_1$$

$$\sum_{i=1}^{n} a_i b_i \geq a_1 b_3 + a_2 b_4 + a_3 b_5 + \cdots + a_n b_2$$

....

$$\sum_{i=1}^{n} a_i b_i \geq a_1 b_n + a_2 b_1 + a_3 b_2 + \cdots + a_n b_{n-1}$$

By adding these inequalities, the inequality

$$n\sum_{i=1}^{n} a_i b_i \geq a_1(b_1 + b_2 + \cdots + b_n) + a_2(b_1 + b_2 + \cdots + b_n)$$
$$+ \cdots + a_n(b_1 + b_2 + \cdots + b_n)$$

is obtained, which, after taking out the common factor $(b_1 + b_2 + \cdots + b_n)$, leads to the Chebyshev's inequality (2.17). For sequences $a_1 \geq, \ldots, \geq a_n$ and $b_1 \leq, \ldots, \leq b_n$, the Chebyshev's sum inequality (2.18) can be proved in a similar fashion, by using the rearrangement inequality.

Chebyshev's sum inequality provides the unique opportunity to segment an initial complex problem into simple problems. The complex terms $a_i b_i$ in inequalities (2.17) and (2.18) are segmented into simpler terms involving a_i and b_i.

The segmentation capability provided by the Chebyshev's sum inequality will be illustrated by evaluating the lower bound of $\sum_{i=1}^{n} x_i^2$ if $x_1 + x_2 + \cdots + x_n = 1$.

Without loss of generality, it can be assumed that $x_1 \geq x_2 \geq \ldots \geq x_n$. By setting $a_1 = x_1, a_2 = x_2, \ldots, a_n = x_n$ and $b_1 = x_1, b_2 = x_2, \ldots, b_n = x_n$, the conditions for the Chebyshev's inequality (2.17) are fulfilled and

$$\frac{x_1^2 + \cdots + x_n^2}{n} \geq \frac{x_1 + \cdots + x_n}{n} \cdot \frac{x_1 + \cdots + x_n}{n} \tag{2.19}$$

Substituting $x_1 + x_2 + ... + x_n = 1$ in inequality (2.19) gives the lower bound of $\sum_{i=1}^{n} x_i^2$:

$$\sum_{i=1}^{n} x_i^2 \geq 1/n \tag{2.20}$$

2.1.7 Muirhead's Inequality

Consider the two non-increasing sequences $a_1 \geq a_2 \geq, \ldots, \geq a_n$ and $b_1 \geq b_2 \geq, \ldots, \geq b_n$ of positive real numbers. The sequence $\{a\}$ is said to majorise the sequence $\{b\}$ if the following conditions are fulfilled:

$$a_1 \geq b_1;\ a_1 + a_2 \geq b_1 + b_2; \ldots;\ a_1 + a_2 + \cdots + a_{n-1} \geq b_1 + b_2 + \cdots + b_{n-1};$$

$$a_1 + a_2 + \cdots + a_{n-1} + a_n = b_1 + b_2 + \cdots + b_{n-1} + b_n \tag{2.21}$$

If the sequence $\{a\}$ majorises the sequence $\{b\}$ and $x_1, x_2, \ldots, x_n$ are non-negative, the *Muirhead's inequality*

$$\sum_{\text{sym}} x_1^{a_1} x_2^{a_2} \ldots x_n^{a_n} \geq \sum_{\text{sym}} x_1^{b_1} x_2^{b_2} \ldots x_n^{b_n} \tag{2.22}$$

holds (Hardy et al., 1999).

For any set of non-negative numbers $x_1, x_2, \ldots, x_n$, the symmetric sum $\sum_{\text{sym}} x_1^{a_1} x_2^{a_2} \ldots x_n^{a_n}$, when expanded, includes $n!$ terms. Each term is formed by a permutation of the elements of the sequence $a_1, a_2, \ldots, a_n$. For example, if $\{a\} = [2,1,0]$, then

$$\sum_{\text{sym}} x_1^2 x_2^1 x_3^0 = x_1^2 x_2 + x_1^2 x_3 + x_2^2 x_1 + x_2^2 x_3 + x_3^2 x_1 + x_3^2 x_2$$

If $\{a\} = [2,0,0]$, then

$$\sum_{\text{sym}} x_1^2 x_2^0 x_3^0 = 2x_1^2 + 2x_2^2 + 2x_3^2$$

Here is an application example featuring an inequality that follows directly from the Muirhead's inequality (2.22). Consider a set of real, non-negative numbers x_1, x_2, x_3. It can be shown that the next inequality holds:

$$x_1^4 + x_2^4 + x_3^4 \geq x_1^2 x_2 x_3 + x_2^2 x_3 x_1 + x_3^2 x_1 x_2 \tag{2.23}$$

Consider the set of non-negative numbers x_1, x_2, x_3 and the sequences $\{a\} = [4,0,0]$ and $\{b\} = [2,1,1]$. Clearly, the sequence $\{a\} = [4,0,0]$ majorises the sequence $\{b\} = [2,1,1]$ because the conditions (2.21) are fulfilled:

$$4 > 2;\ 4+0 > 2+1;\ \text{and } 4+0+0 \geq 2+1+1.$$

According to the Muirhead's inequality (2.22),

$$2\times\left(x_1^4 + x_2^4 + x_3^4\right) \geq 2\left(x_1^2 x_2 x_3 + x_2^2 x_3 x_1 + x_3^2 x_1 x_2\right)$$

holds, which implies inequality (2.23).

2.2 ALGEBRAIC INEQUALITIES THAT PERMIT NATURAL MEANINGFUL INTERPRETATION

2.2.1 Symmetric Algebraic Inequalities Whose Terms Can Be Interpreted as Probabilities

Some symmetric algebraic inequalities permit natural interpretation as probabilities if particular constraints are imposed on the variables of the inequality. Such is, for example, the wide class of symmetric algebraic inequalities with imposed additional constraints. For example, the inequalities

$$x^2 + y^2 + z^2 \geq 1/3 \tag{2.24}$$

$$2xy + 2yz + 2zx \leq 2/3 \tag{2.25}$$

with imposed constraints $0 < x < 1$, $0 < y < 1$, $0 < z < 1$ and $x + y + z = 1$, admit natural interpretation of x, y and z as fractions of items of varieties X, Y and Z. Inequality (2.24) is a special case of inequality (2.20) for $n = 3$. Inequality (2.25) can be obtained from the identity $(x + y + z)^2 = x^2 + y^2 + z^2 + 2xy + 2yz + 2zx = 1$ and inequality (2.24). Indeed,

$$2xy + 2yz + 2zx = 1 - (x^2 + y^2 + z^2) \leq 1 - 1/3 = 2/3$$

The left-hand side of inequality (2.24) is a sum of three terms which implies the application of *the total probability theorem* (DeGroot, 1989) in determining the probability of a compound event which includes three mutually exclusive events.

Such a compound event for inequality (2.24) is, for example, selecting two items of the same variety from a large batch containing items of three varieties X, Y and Z. The corresponding fractions of items in the batch, from the separate varieties, are x, y and z ($0 < x < 1, 0 < y < 1, 0 < z < 1$ and $x + y + z = 1$). Selecting two items of variety X, variety Y or variety Z are mutually exclusive events, and

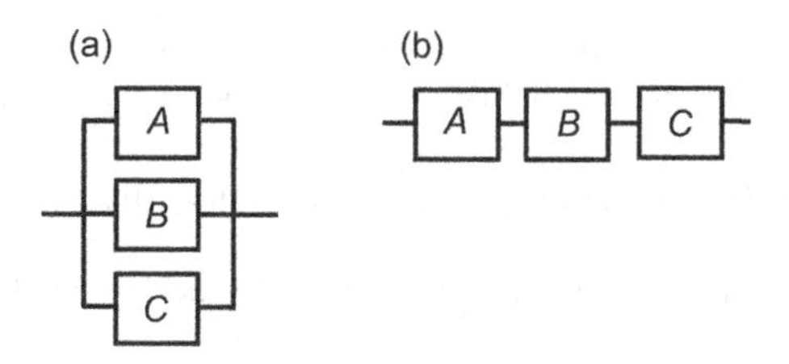

FIGURE 2.2 Three components A, B and C logically arranged (a) in parallel and (b) in series.

according to the total probability theorem, their probabilities are added to obtain the probability of selecting two items of the same variety.

Similarly, the left-hand side of inequality (2.25) can be interpreted as the probability of selecting two items from two different varieties. Selecting two items of varieties XY, YZ or ZX are mutually exclusive events, and according to the total probability theorem, their probabilities $2xy$, $2yz$ and $2zx$ are added.

Symmetric inequalities similar to

$$(1-a)(1-b)(1-c) \leq 1-abc \tag{2.26}$$

where $0 \leq a \leq 1$, $0 \leq b \leq 1$ and $0 \leq c \leq 1$ can be interpreted easily if a, b and c are chosen to denote the reliabilities of components A, B and C, correspondingly. The reliabilities a, b and c are effectively the probabilities that the components will be in working condition. The probabilities that the components will be in a failed condition are $1-a$, $1-b$ and $1-c$, correspondingly.

The system in Figure 2.2a, where the components A, B and C have been logically arranged in parallel, is in failed state only if all components are in failed state, while the system in Figure 2.2b, where all components are logically arranged in series, is in working state if all components are working. Consequently, the left-hand side of inequality (2.26) can be interpreted as probability of failure of the system in Figure 2.2a, and the right-hand side of the inequality can be interpreted as probability of failure of the system in Figure 2.2b.

Inequality (2.26) predicts that the probability of failure of the system in Figure 2.2a is smaller than the probability of failure of the system in Figure 2.2b.

2.2.2 Transforming Algebraic Inequalities to Make Them Interpretable

By appropriate transformations, some symmetric inequalities can be made interpretable.

A common transformation technique is to multiply both sides of the inequality by a constant. Thus, multiplying both sides by the constant (1/2), for example, creates the possibility to interprete terms of type a^2 / b as elastic strain energy or stored electric energy (see Chapter 4 for details). Multiplying both sides of an inequality by $1/n$, where n is the number of terms in the left- or right-hand side of the inequality, creates the possibility to apply the total probability theorem and interpret the inequality as a probability of a compound event composed of mutually exclusive events.

Indeed, if event B can occur with any of the n mutually exclusive and exhaustive events A_i $\left(P(A_i \cap A_j) = 0 \text{ and } \sum_{i=1}^{n} P(A_i) = 1 \right)$, according to the total probability theorem, the probability $P(B)$ of event B is given by:

$$P(B) = P(B \mid A_1)P(A_1) + P(B \mid A_2)P(A_2) + \cdots + P(B \mid A_n)P(A_n) \quad (2.27)$$

Now if events A_i have the same chance of occurring, the probability $P(A_i)$ of their occurrence is equal to $1/n$. The probability $P(B)$ of event B then becomes

$$P(B) = (1/n)P(B \mid A_1) + (1/n)P(B \mid A_2) + \cdots + (1/n)P(B \mid A_n) \quad (2.28)$$

Inequality (2.15) that has been proved in Section 2.1.4 is one such example.

Multiplying both sides of inequality (2.15) by $1/3$ gives the inequality

$$(1/3)x^2 + (1/3)y^2 + (1/3)z^2 \geq (1/3)xy + (1/3)yz + (1/3)zx \quad (2.29)$$

If x, y and z stand for the fractions of high-reliability components characterising three suppliers X, Y and Z, the left-hand side of inequality (2.29) can be interpreted as probability of purchasing two high-reliability components from a randomly selected supplier. Indeed, the term $(1/3)x^2$ in inequality (2.29) is the probability that the first supplier will be randomly selected and both components purchased from this supplier will be high-reliability components. The term $(1/3)y^2$ is the probability that the second supplier will be randomly selected and both components purchased from this supplier will be high-reliability components. Finally, the term $(1/3)z^2$ is the probability that the third supplier will be randomly selected and both components purchased from this supplier will be high-reliability components. These events are mutually exclusive and exhaustive, and according to the total probability theorem, $p_1 = (1/3)x^2 + (1/3)y^2 + (1/3)z^2$ is the probability of purchasing two high-reliability components from a randomly selected supplier. With a similar reasoning, it can be shown that the right-hand side of inequality (2.29), $p_2 = (1/3)xy + (1/3)yz + (1/3)zx$, can be interpreted as probability of purchasing two high-reliability components from two randomly selected suppliers.

The next example of transformations applied to symmetric inequalities is related to the inequality

$$a^2 + b^2 \geq 2ab \quad (2.30)$$

where $a, b > 0$. The transformation consists of dividing both sides of the inequality by the positive quantity $a + b$. This operation does not change the direction of the inequality, but the terms $x = a/(a+b)$ and $y = b/(a+b)$ permit interpreting them as fractions of components from two varieties X and Y, correspondingly, in a large batch of components. The original inequality is reduced to

$$x^2 + y^2 \geq 2xy \tag{2.31}$$

where $x, y > 0$ and $x + y = 1$. The left-hand side of inequality (2.31) can now be interpreted as the probability of selecting two items of the same variety from a large batch, and the right-hand side of the inequality can be interpreted as the probability of selecting two items of different variety from the same batch.

2.2.3 Inequalities Based on Sub- and Super-Additive Functions

This is a very important class of algebraic inequalities that provides the possibility for segmentation of a controlling factor. Inequalities based on sub- and super-additive functions have a number of key potential applications in increasing the effect of additive quantities (factors).

Extensive quantities are typical additive quantities. They change with changing the size of their supporting objects/systems (DeVoe, 2012; Mannaerts, 2014). DeVoe (2012), for example, introduced 'extensivity' test which consists of dividing the system by an imaginary surface into two parts. Any quantity characterising the system that is the sum of the same quantity characterising the two parts is an *extensive* quantity, and any quantity that has the same value in each part of the system is an *intensive* quantity.

Examples of extensive quantities are mass, weight, amount of substance, number of particles, volume, distance, energy (kinetic energy, gravitational energy, electric energy, elastic energy, surface energy, internal energy), work, power, heat, force, momentum, electric charge, electric current, heat capacity, electric capacity, resistance (when the elements are connected in series), enthalpy and fluid flow.

Intensive quantities characterise the object/system locally and do not change with changing the size of the supporting objects/systems. Examples of intensive properties are, temperature, pressure, density, concentration, hardness, velocity and surface tension.

In the Euclidean space of one dimension, the sub-additive function $f(x)$ satisfies the inequality

$$f(x_1 + x_2) \leq f(x_1) + f(x_2) \tag{2.32}$$

for any pair of points x_1 and x_2 in the domain of definition.

If the direction of the inequality is reversed, the function $f(x)$ is super-additive:

$$f(x_1 + x_2) \geq f(x_1) + f(x_2) \tag{2.33}$$

In the Euclidean space of two dimensions, the multivariable sub-additive function $f(x, y)$ satisfies the inequality

$$f(x_1 + x_2,\ y_1 + y_2) \le f(x_1, y_1) + f(x_2, y_2) \tag{2.34}$$

for any pair of points (x_1, y_1) and (x_2, y_2) in the domain of definition.

If the direction of the inequality is reversed, the function $f(x,y)$ is super-additive:

$$f(x_1 + x_2,\ y_1 + y_2) \ge f(x_1, y_1) + f(x_2, y_2) \tag{2.35}$$

Consider n points with positive coordinates x_i in the definition domain of the function *f()*. By induction, from definition (2.32), it follows that

$$f(x_1 + x_2 + \cdots + x_n) \le f(x_1) + f(x_2) + \cdots + f(x_n) \tag{2.36}$$

while from definition (2.33), it follows that

$$f(x_1 + x_2 + \cdots + x_n) \ge f(x_1) + f(x_2) + \cdots + f(x_n) \tag{2.37}$$

Key results related to sub- and super-additive functions of a single variable have been stated in Alsina and Nelsen (2010). Thus, if a function $f(x)$, with a domain $[0,\infty)$ and range $[0,\infty)$, is concave and $f(0) \ge 0$, then the function is sub-additive: $f(x_1 + x_2 + \cdots + x_n) \le f(x_1) + f(x_2) + \cdots + f(x_n)$. If the function $f(x)$ is convex and $f(0) \le 0$, then it is super-additive: $f(x_1 + x_2 + \cdots + x_n) \ge f(x_1) + f(x_2) + \cdots + f(x_n)$.

Consider now n pair of points (a_i, b_i) with positive coordinates a_i, b_i from the definition domain of the function $f(x,y)$. From the definition (2.34), it follows that

$$f(a_1 + a_2 + \cdots + a_n,\ b_1 + b_2 + \cdots + b_n) \le f(a_1, b_1) + f(a_2, b_2) + \cdots + f(a_n, b_n) \tag{2.38}$$

while from the definition (2.35), it follows that

$$f(a_1 + a_2 + \cdots + a_n,\ b_1 + b_2 + \cdots + b_n) \ge f(a_1, b_1) + f(a_2, b_2) + \cdots + f(a_n, b_n) \tag{2.39}$$

Relationships (2.36)–(2.39) can be obtained easily by mathematical induction and derivation details have been omitted.

Inequalities (2.38) and (2.39) do not change their direction upon any permutation of a_i and b_i. They have a number of powerful potential applications in increasing/decreasing the effect of additive quantities (factors).

Inequalities (2.36) and (2.39) can, for example, be applied for optimising processes. If the function $f(x)$ reflects the effect/output of an additive controlling factor x and x_i ($i = 1, \ldots, n$) denotes a smaller part of the factor, inequalities (2.36) and (2.37) provide the unique opportunity to increase the effect of the factor by segmenting it or aggregating it, depending on whether the function $f(x)$ is concave or convex. If the function is concave, with a domain $[0,\infty)$ and range $[0,\infty)$, segmenting the factor results in a larger output. Conversely, if the function is convex in this domain and if the function is zero when the factor is zero, aggregating the factor results in a larger output.

An important condition for using inequalities (2.36) and (2.37) is the outputs $f(x_1), f(x_2),\ldots, f(x_n)$ after segmenting the controlling factor x to be also additive. This is fulfilled if the outputs $f(x_1), f(x_2),\ldots, f(x_n)$ are additive quantities such as energy, power, force, damage, profit, pollution, mass, number and volume.

In order to apply inequalities (2.38) or (2.39), the variables a_i, b_i and the output quantities $f(a_i, b_i)$ must all represent additive quantities. Inequality (2.38) effectively states that the effect of the additive quantities $a = \sum_{i=1}^{n} a_i$ and $b = \sum_{i=1}^{n} b_i$ can be increased by segmenting them into smaller parts a_i and b_i, $i = 1,\ldots,n$ and accumulating their effects $f(a_i, b_i)$ (represented by the sum of the terms on the right-hand side of inequality (2.38). Similarly, inequality (2.39) effectively states that the effect of the additive quantities $a = \sum_{i=1}^{n} a_i$ and $b = \sum_{i=1}^{n} b_i$ can be decreased by segmenting them into smaller parts a_i and b_i, $i = 1,\ldots,n$ and accumulating their effects $f(a_i, b_i)$ represented by the sum of the terms on the right-hand side of inequality (2.39).

Inequalities (2.38) and (2.39) have a universal application in science and technology as long as a_i, b_i and the terms $f(a_i, b_i)$ are additive quantities and have meaningful interpretation.

Inequalities similar to (2.38) and (2.39) can be derived for any number of factors by using the definition of sub-/super-additive functions. For example, for three factors a, b and c, the inequality

$$\begin{aligned} &f(a_1 + \cdots + a_n,\, b_1 + \cdots + b_n, c_1 + \cdots + c_n) \\ &\leq f(a_1, b_1, c_1) + f(a_2, b_2, c_2) + \cdots + f(a_n, b_n, c_n) \end{aligned} \tag{2.40}$$

is applicable for a sub-additive function $f(a,b,c)$ and the inequality

$$\begin{aligned} &f(a_1 + \cdots + a_n,\, b_1 + \cdots + b_n, c_1 + \cdots + c_n) \\ &\geq f(a_1, b_1, c_1) + f(a_2, b_2, c_2) + \cdots + f(a_n, b_n, c_n) \end{aligned} \tag{2.41}$$

is applicable for a super-additive function $f(a,b,c)$.

Inequalities based on sub- and super-additive functions are important classes of algebraic inequalities that permit easy meaningful interpretation. The knowledge derived from interpreting sub- and super-additive inequalities can be used to obtain superior performance from a system or process.

Inequalities based on single- and multivariable sub- or super-additive functions can always be interpreted if the variables and the separate terms in the inequalities represent additive quantities.

Sufficient conditions for sub-additivity and super-additivity of a single-variable function can also be stated. If a function $f(x)$, with a domain $[0,\infty)$ and range $[0,\infty)$, is a concave single-variable function (see Section 2.1.3 for a definition), then the function is sub-additive $\left(f(x_1+x_2+\cdots+x_n)\le f(x_1)+f(x_2)+\cdots+f(x_n)\right)$. If the function $f(x)$ is convex and $f(0)\le 0$, then the function is super-additive $\left(f(x_1+x_2+...+x_n)\ge f(x_1)+f(x_2)+\cdots+f(x_n)\right)$.

2.2.3.1 Proof

The correctness of the first statement can be proved by an argument based on the fact that $f(x)$ is a concave function. Note that $w_{1,k}=x_k\Big/\sum_{i=1}^{n}x_i$ and $w_{2,k}=\left(\sum_{i=1}^{n}x_i-x_k\right)\Big/\sum_{i=1}^{n}x_i$ can be treated as weights because $0\le w_{1,k}\le 1$, $0\le w_{2,k}\le 1$ and $w_{1,k}+w_{2,k}=1$. Because $f(x)$ is a concave function, for the values $x=x_1+x_2+\cdots+x_n$ and $x=0$, the Jensen's inequality for concave functions gives:

$$f(x_1)=f\left(\frac{x_1}{\sum_{i=1}^{n}x_i}\times\sum_{i=1}^{n}x_i+\frac{\sum_{i=1}^{n}x_i-x_1}{\sum_{i=1}^{n}x_i}\times 0\right)\ge\frac{x_1}{\sum_{i=1}^{n}x_i}f(x_1+\cdots+x_n)$$

$$+\frac{\sum_{i=1}^{n}x_i-x_1}{\sum_{i=1}^{n}x_i}f(0) \tag{2.42}$$

$$f(x_2)=f\left(\frac{x_2}{\sum_{i=1}^{n}x_i}\times\sum_{i=1}^{n}x_i+\frac{\sum_{i=1}^{n}x_i-x_2}{\sum_{i=1}^{n}x_i}\times 0\right)\ge\frac{x_2}{\sum_{i=1}^{n}x_i}f(x_1+\cdots+x_n)$$

$$+\frac{\sum_{i=1}^{n}x_i-x_2}{\sum_{i=1}^{n}x_i}f(0) \tag{2.43}$$

....

$$f(x_n) = f\left(\frac{x_n}{\sum_{i=1}^{n} x_i} \times \sum_{i=1}^{n} x_i + \frac{\sum_{i=1}^{n} x_i - x_n}{\sum_{i=1}^{n} x_i} \times 0 \right) \geq \frac{x_n}{\sum_{i=1}^{n} x_i} f(x_1 + \cdots + x_n)$$

$$+ \frac{\sum_{i=1}^{n} x_i - x_n}{\sum_{i=1}^{n} x_i} f(0) \qquad (2.44)$$

Adding all n inequalities (2.42)–(2.44) yields

$$f(x_1) + f(x_2) + \cdots + f(x_n) \geq f(x_1 + x_2 + \cdots + x_n) + (n-1)f(0) \qquad (2.45)$$

Since $f(0) \geq 0$ (the range of the function $f(x)$ is $[0, \infty)$),

$$f(x_1) + f(x_2) + \cdots + f(x_n) \geq f(x_1 + x_2 + \cdots + x_n) + (n-1)f(0) \geq f(x_1 + x_2 + \cdots + x_n)$$

This completes the proof of inequality (2.36).

Note that if $f(0) > 0$, equality in (2.36) cannot be attained:

$$f(x_1) + f(x_2) + \cdots + f(x_n) \geq f(x_1 + x_2 + \cdots + x_n) + (n-1)f(0) > f(x_1 + x_2 + \cdots + x_n)$$

In the same way, inequality (2.37) can also be proved if the function $f(x)$ is convex and $f(x) \leq 0$. These are sufficient conditions for super-additivity of a single-variable function.

Inequalities can also be based on multivariable sub- and super-additive functions. Multivariate sub-additive functions have been discussed in Rosenbaum (1950).

A special case of the general super-additive function (2.39) is the algebraic inequality

$$\sqrt{ab} \geq \sqrt{a_1 b_1} + \sqrt{a_2 b_2} + \cdots + \sqrt{a_n b_n} \qquad (2.46)$$

where both controlling factors $a = a_1 + a_2 + \cdots + a_n$ and $b = b_1 + b_2 + \cdots + b_n$ are additive quantities.

The role of the function $f(a, b)$ in inequality (2.39) is played by the function $f(a, b) \equiv \sqrt{ab}$ in inequality (2.46).

Inequality (2.46) can be derived from the Cauchy–Schwarz inequality by making the substitutions:

$x_1 = \sqrt{a_1}$, $x_2 = \sqrt{a_2}$,…,$x_n = \sqrt{a_n}$; $y_1 = \sqrt{b_1}$, $y_2 = \sqrt{b_2}$,…,$y_n = \sqrt{b_n}$. Applying the Cauchy–Schwarz inequality to the sequences $x_1, x_2,\ldots, x_n$ and $y_1, y_2,\ldots, y_n$ yields inequality (2.46):

$$\sqrt{a_1 b_1} + \sqrt{a_2 b_2} + \cdots + \sqrt{a_n b_n} \le$$

$$\sqrt{\left[\left(\sqrt{a_1}\right)^2 + \cdots + \left(\sqrt{a_n}\right)^2\right] \times \left[\left(\sqrt{b_1}\right)^2 + \cdots + \left(\sqrt{b_n}\right)^2\right]}$$

Relevant meaning can be created for the additive quantities a_i, b_i entering inequality (2.46). The condition for meaningfully interpreting inequality (2.46) is the quantity represented by $\sqrt{a_i b_i}$ to be also an additive quantity.

2.2.4 Bergström Inequality and Its Natural Interpretation

Often transformations of basic algebraic inequalities are a necessary prerequisite for their meaningful interpretation. Such a transformation will be demonstrated with the standard Cauchy–Schwarz inequality (2.1). If the substitutions $a_i = \dfrac{x_i}{\sqrt{y_i}}$ $(i = 1,\ldots,n)$ and $b_i = \sqrt{y_i}$ $(i = 1,\ldots,n)$ are made in the Cauchy–Schwarz inequality (2.1):

$$\left(\frac{x_1}{\sqrt{y_1}}\sqrt{y_1} + \frac{x_2}{\sqrt{y_2}}\sqrt{y_2} + \cdots + \frac{x_n}{\sqrt{y_n}}\sqrt{y_n}\right)^2 \le \left(\left(x_1/\sqrt{y_1}\right)^2 + \cdots + \left(x_n/\sqrt{y_n}\right)^2\right)$$
$$\left(\left(\sqrt{y_1}\right)^2 + \cdots + \left(\sqrt{y_n}\right)^2\right)$$

the result is the Bergström inequality (Pop, 2009; Sedrakyan and Sedrakyan, 2010)

$$\frac{x_1^2}{y_1} + \frac{x_2^2}{y_2} + \cdots + \frac{x_n^2}{y_n} \ge \frac{(x_1 + x_2 + \cdots + x_n)^2}{y_1 + y_2 + \cdots + y_n} \tag{2.47}$$

which is effectively the modified Cauchy–Schwarz inequality. Inequality (2.48) is valid for any sequence $x_1, x_2,\ldots, x_n$ of real numbers and any sequence $y_1, y_2,\ldots, y_n$ of positive real numbers. Equality in (2.47) is attained only if $x_1 / y_1 = x_2 / y_2 = \ldots = x_n / y_n$.

In order to apply inequality (2.47) to additive quantities, the quantities x_i, y_i and the ratios x_i^2 / y_i must all be additive quantities. In this case, the inequality provides mechanism for increasing the effect of the quantities $x = \left(\sum_{i=1}^{n} x_i\right)^2$ and

$y = \sum_{i=1}^{n} y_i$ by segmenting them into smaller quantities x_i and y_i, $i = 1,\ldots,n$ and accumulating the individual effects x_i^2 / y_i.

Relevant meaning can be created for the quantities x_i^2, y_i entering inequality (2.47). The quantity x^2 could, for example, be the square of force or voltage and the quantity b could, for example, be area or resistance (of elements arranged in series). A necessary condition for interpreting inequality (2.47) is the quantity represented by x_i^2 / y_i to be an additive quantity. Thus, the strain energy of a bar subjected to tension is proportional to Force^2/Area and the strain energy is an additive quantity. The quantity Voltage^2/Resistance equals dissipated power which is also an additive quantity.

2.2.5 A New Algebraic Inequality Which Provides Possibility for a Segmentation of Additive Factors

Consider the general inequality

$$f(a_1 + a_2 + \cdots + a_n,\ b_1 + b_2 + \cdots + b_n) \le k\,[f(a_1, b_1) + f(a_2, b_2) + \cdots + f(a_n, b_n)] \tag{2.48}$$

where k is a constant.

Consider also the next algebraic inequality

$$\frac{a_1}{x_1} + \frac{a_2}{x_2} + \cdots + \frac{a_n}{x_n} \ge n\frac{a_1 + a_2 + \cdots + a_n}{x_1 + x_2 + \cdots + x_n} \tag{2.49}$$

where $a_1 \ge a_2 \ge \ldots \ge a_n$ and $x_1 \le x_2 \le \ldots \le x_n$ are positive additive quantities. Note that inequality (2.47) is a special case of inequality (2.48) when k=1, and inequality (2.49) is also a special case of inequality (2.48) when k=1/n. Equality in inequality (2.49) is attained if $a_i / x_i = a_j / x_j$ or if $x_i = x_j$, (i=1,...,n; j=1,...,n).

This inequality can be proved by applying the Chebyshev's sum inequality, the AM–GM inequality and the technique 'strengthening of an inequality'.

Because $x_1 \le x_2 \le \ldots \le x_n$, the ranking $1/x_1 \ge 1/x_2 \ge \ldots \ge 1/x_n$ holds and the Chebyshev's sum inequality gives:

$$n\left(\frac{a_1}{x_1} + \frac{a_2}{x_2} + \cdots + \frac{a_n}{x_n}\right) \ge (a_1 + a_2 + \cdots + a_n) \times (1/x_1 + 1/x_2 + \cdots + 1/x_n) \tag{2.50}$$

From the AM–GM inequality,

$$1/x_1 + 1/x_2 + \cdots + 1/x_n \ge \frac{n}{\sqrt[n]{x_1 x_2 \ldots x_n}}; \tag{2.51}$$

therefore,

$$n\left(\frac{a_1}{x_1} + \frac{a_2}{x_2} + \cdots + \frac{a_n}{x_n}\right) \ge \frac{n(a_1 + a_2 + \cdots + a_n)}{\sqrt[n]{x_1 x_2 \ldots x_n}} \tag{2.52}$$

From the AM–GM inequality,

$$\sqrt[n]{x_1 x_2 \ldots x_n} \le \frac{x_1 + x_2 + \cdots + x_n}{n} \tag{2.53}$$

holds, and substituting $\frac{x_1 + x_2 + \cdots + x_n}{n}$ in inequality (2.52) instead of $\sqrt[n]{x_1 x_2 \ldots x_n}$, only strengthens inequality (2.52). The result is the inequality

$$n\left(\frac{a_1}{x_1} + \frac{a_2}{x_2} + \cdots + \frac{a_n}{x_n}\right) \ge \frac{n^2 (a_1 + a_2 + \cdots + a_n)}{x_1 + x_2 + \cdots + x_n} \tag{2.54}$$

which, after the division of both sides by n, gives inequality (2.49).

Inequality (2.49) provides a mechanism for increasing at least n times the effect of the additive quantities $a = \sum_{i=1}^{n} a_i$ and $x = \sum_{i=1}^{n} x_i$ by segmenting them into smaller parts a_i, x_i, $i = 1,\ldots,n$ and accumulating their individual effects a_i / x_i. In order to apply inequality (2.49), the ratio a_i / x_i of the additive quantities a_i and x_i must also be an additive quantity.

2.3 TESTING ALGEBRAIC INEQUALITIES BY MONTE CARLO SIMULATION

Before attempting to prove a derived or conjectured inequality rigorously, it is important to confirm it first by testing. An attempt is made to prove the inequality rigorously only if the testing provides support for the conjectured inequality. No attempt is made to prove the inequality rigorously if, during the testing, the inequality has been falsified by a counterexample.

Suppose that n distinct components with unknown reliabilities $a_1, a_2, \ldots, a_n$ are used for building two system configurations, with reliabilities given by the functions $f(a_1,\ldots,a_n)$ and $g(a_1,\ldots,a_n)$. Consider a case where it is necessary to test the conjecture that the system configuration with reliability $f(a_1,\ldots,a_n)$ is intrinsically superior to the system configuration with reliability $g(a_1,\ldots,a_n)$, irrespective of the reliabilities $a_1, a_2, \ldots, a_n$ of the components building the systems. To perform this test, it suffices to test that the conjectured inequality $f(a_1,\ldots,a_n) > g(a_1,\ldots,a_n)$ is not contradicted during multiple simulation trials involving random combinations of values for the unknown reliabilities $a_1, a_2, \ldots, a_n$. This can be done by running the Monte Carlo simulation Algorithm 2.1 given in pseudo-code.

In general, the variables a_i entering the conjectured inequality $f(a_1,\ldots,a_n) > g(a_1,\ldots,a_n)$, vary within specified intervals $L_i \le a_i \le U_i$, $i = 1,\ldots,n$. If a_i are reliabilities of components, then $L_i = 0$, $U_i = 1$.

The essence of Algorithm 2.1 for testing a conjectured inequality, is repeated sampling from the intervals of variation of each variable a_i entering the inequality, substituting the sampled values in the inequality and checking whether the inequality $f(a_1,\ldots,a_n) > g(a_1,\ldots,a_n)$ holds. Even a single combination of values for the variable a_i, for which the inequality does not hold, disproves the inequality and shows that there is no point in attempting a rigorous proof because a counterexample had been found that falsified the conjectured inequality.

If the inequality holds for millions of generated random values for the sampled variables, a strong support is obtained for the conjecture that the tested inequality is true. Such support, however, cannot replace a rigorous proof. The inequality must still be proved rigorously by using some of the techniques for proving inequalities.

The algorithm in pseudo-code, for confirming or disproving an inequality by a Monte Carlo simulation, is given next.

Algorithm 2.1

```
a=[];
L=[L1,L2,...,Ln];   U=[U1,U2,...,Un];

flag=0;
for k=1 to num_trials do
{
 for k=1 to n do  a[k]=L[k]+(U[k] - L[k]) x rand();

 y1=f(a[1],a[2],...,a[n]);
 y2=g(a[1],a[2],...,a[n]);
 y=y1-y2;

 if(y<0) then { flag=1; break;}

}

if (flag==0) then print('The tested inequality has never
been contradicted');
else print('The tested inequality has been disproved');
```

Initially, in the loop

```
for k=1 to n do  a[k]=L[k]+(U[k]-L[k]) x rand(),
```

random values are assigned to the variables a[k], (k=1,...,n), entering the inequality.

This is done by calling the function **rand**(), which returns a random number uniformly distributed in the interval (0,1). The value returned by **rand**() is transformed linearly by the statement `a[k]=L[k]+(U[k]-L[k]) x rand()` into a random value a[k] uniformly distributed in the interval (L[k], U[k]). L[k] and U[k] are the lower and upper limit of the range for variable a[k].

Next, the left- and right-hand sides of the tested inequality are evaluated by calling the functions f() and g() with the statements

```
y1=f(a[1],a[2],...,a[n]); y2=g(a[1],a[2],...,a[n]);
```

The values returned by the functions are stored in the variables y1 and y2.

The difference y=y1-y2 is then checked for being smaller than zero. A difference smaller than zero corresponds to a case where the tested inequality has been contradicted. In the statement

```
if(y<0) then { flag=1; break;}
```

if the tested inequality has been contradicted, a variable serving as flag is assigned value equal to one (`flag=1`) and the simulation loop is exited immediately with the statement '`break`' because a counterexample has been found. At the end of the algorithm, the content of the variable '`flag`' is checked. If the flag remained zero during the simulations, this is an indication that the tested inequality has never been contradicted during the simulation trials. This indicates that the tested inequality is probably true and a rigorous proof can be attempted. If, during the simulations, the variable '`flag`' changed its value to one, this is an indication that there had been a combination of values for the variables entering the tested inequality for which the inequality does not hold. This means that there is no point in searching for rigorous proof because a counterexample has been found which disproved the conjectured inequality.

Consider the non-trivial algebraic inequality:

$$(1-a^2b)(1-b^2c)(1-c^2a) \geq (1-a^3)(1-b^3)(1-c^3)$$

where a,b,c are positive real numbers smaller than 1. Before an attempt can be made to prove this inequality rigorously, the inequality can be tested to confirm its validity for various combinations of values for the variables a, b and c.

This can be done with the next algorithm.

Algorithm 2.2

```
tmp=0;
num_trials=10000000;

flag=0;

for k=1 to num_trials do
{
 a=rand(); b=rand(); c=rand();

 y1=(1-a^2*b)*(1-b^2*c)*(1-c^2*a);
 y2=(1-a^3)*(1-b^3)*(1-c^3);

 y=y1-y2;

 if(y<0)
   {
```

```
      flag=1;
      break;
    }
}
if (flag==0)
 print('The inequality has never been contradicted')
 else
  print('The inequality has been contradicted');
```

This algorithm has actually been implemented and the result from 10 million simulation trials was that the value of the flag had never been changed to 1, which led to the output '`The inequality has never been contradicted`'. The absence of a contradiction for 10 million random combinations of values for the variables a, b and c is a strong indication that the conjectured inequality is probably true. As a result, an attempt for a rigorous proof of the inequality can be made. The described Monte-Carlo simulation algorithm 2.1 was used to validate all algebraic inequalities presented in the book.

3 Generating Knowledge about Physical Systems by Meaningful Interpretation of Algebraic Inequalities

The inverse approach related to generating knowledge by interpreting an existing algebraic inequality has been outlined in Chapter 1. Attaching meaning to the variables in the abstract inequality combined with meaningful interpretation of the different parts of the inequality links the abstract inequality with reality. As a result of the interpretation, often, the abstract inequality expresses a new physical property. Suppose that the left hand side of the inequality is interpreted as a particular performance output of a particular system and the right-hand side is interpreted as the same performance output characterising a system built with the same components but in a different configuration. The interpretation of the inequality then ranks the performance of the two competing system configurations.

3.1 AN ALGEBRAIC INEQUALITY RELATED TO EQUIVALENT PROPERTIES OF ELEMENTS CONNECTED IN SERIES AND PARALLEL

Consider an example of this approach which starts from the correct algebraic inequality

$$(x_1 + x_2 + \cdots + x_n) \geq n^2 \left(\frac{1}{1/x_1 + 1/x_2 + \cdots + 1/x_n} \right) \tag{3.1}$$

valid for any set of n non-negative quantities x_i. Inequality (3.1) is equivalent to the inequality:

$$\frac{x_1 + x_2 + \cdots + x_n}{n} \geq \frac{n}{1/x_1 + 1/x_2 + \cdots + 1/x_n}$$

which is the Arithmetic-mean–Harmonic mean (AM–HM) inequality introduced in Chapter 2.

DOI: 10.1201/9781003199830-3

Inequality (3.1) can also be proved rigorously by reducing it to the standard Cauchy–Schwarz inequality, which states that for any two sequences of real numbers $\{a_1,a_2,\ldots,a_n\}$ and $\{b_1,b_2,\ldots,b_n\}$, the following inequality holds:

$$(a_1b_1 + a_2b_2 + \cdots + a_nb_n)^2 \leq \left(a_1^2 + a_2^2 + \cdots + a_n^2\right)\left(b_1^2 + b_2^2 + \cdots + b_n^2\right) \quad (3.2)$$

Because the quantities x_i in inequality (3.1) are non-negative, the setting $a_i = \sqrt{x_i}$ and $b_i = 1/\sqrt{x_i}$, $i = 1, n$, can be made. Next, substituting a_i and b_i in the Cauchy–Schwarz inequality (3.2) yields inequality (3.1).

To make inequality (3.1) relevant to a real system, appropriate meaning must be attached to the variables entering the inequality and to its left- and right hand side.

3.1.1 Elastic Components and Resistors Connected in Series and Parallel

Suppose that the variables x_i in inequality (3.1) stand for stiffness of an elastic element i. Now, the two sides of the inequality can be meaningfully interpreted in the following way. The expression $x_1 + x_2 + \cdots + x_n$ on the left-hand side of inequality (3.1) can be interpreted as the equivalent stiffness for n elastic elements connected in parallel (Figure 3.1a). The expression $\dfrac{1}{1/x_1 + 1/x_2 + \cdots + 1/x_n}$ on the right-hand side of the inequality can be interpreted as the equivalent stiffness of n elastic elements connected in series (Figure 3.1b). The inequality then establishes a connection between the equivalent stiffness of n elastic elements connected in parallel (the left-hand side) and the equivalent stiffness of the same n elastic elements connected in series (in the right-hand side). According to the principle of

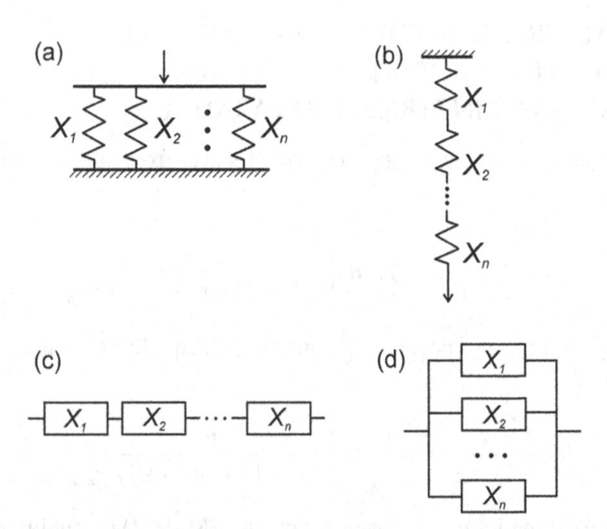

FIGURE 3.1 Interpretations of inequality 3.1 involving (a) elastic elements arranged in parallel; (b) elastic elements arranged in series; (c) resistors arranged in series and (d) resistors arranged in parallel.

non-contradiction, the prediction from inequality (3.1) must be consistent with the properties of the two system configurations. As a result, a new physical property can be inferred from inequality (3.1): the equivalent stiffness of n elastic elements connected in parallel is at least n^2 times larger than the equivalent stiffness of the same elastic elements connected in series, irrespective of the individual stiffness values of the elements (Todinov, 2020b).

The meaning created for the variables and the different parts of inequality (3.1) is not unique and can be altered. Thus, a new meaning can be created if, for example, each x_i stands for 'electrical resistance of element i'.

The expression $x_1 + x_2 + \cdots + x_n$ on the left-hand side of inequality (3.1) can be meaningfully interpreted as the equivalent resistance of a system configuration including n elements connected in series (Figure 3.1c). On the right-hand side of the inequality, the expression $\frac{1}{1/x_1 + 1/x_2 + \cdots + 1/x_n}$ can be meaningfully interpreted as the equivalent resistance of another system configuration including the same n elements connected in parallel (Figure 3.1d). Inequality (3.1) now predicts another physical property: the equivalent resistance of n elements arranged in series is at least n^2 times larger than the equivalent resistance of the same elements arranged in parallel, irrespective of the individual values of the resistances.

This new property can be applied in testing the insulation of high-voltage equipment where the resistances of the insulating elements have extremely large values and are connected in series. Because of the very high resistances of the insulating elements, it is extremely difficult to measure directly not only the equivalent resistance $R_s = x_1 + x_2 + \cdots + x_n$ of n insulating elements connected in series but even the resistance x_i of a single insulating element. If, however, the same insulating elements are connected in parallel (instead of series), it is much easier to measure the equivalent resistance $R_p = 1/(1/x_1 + 1/x_2 + \cdots + 1/x_n)$ of the parallel arrangement because, according to inequality (3.1), it has at least n^2 times smaller magnitude. After the measurement of the equivalent resistance R_p of a parallel arrangement, a conclusion about the equivalent resistance of the same elements connected in series can be made from the inequality $R_s \geq n^2 R_p$. The equivalent resistance R_s of the elements in series will be at least n^2 times larger than the measured equivalent resistance R_p of the same elements in parallel. This is exactly what is needed to assure the safety of the insulation assembly: the resistance of the series arrangement to be larger than a particular critical value.

It needs to be pointed out that for elements of equal resistances, the fact that the equivalent resistance in series is exactly n^2 times larger than the equivalent resistance of the resistors in parallel is a trivial result, known for a long period of time and used for high-resistance standards (Rozhdestvenskaya and Zhutovskii, 1968).

Indeed, for the equivalent resistance of n identical resistors with resistances $x_1 = x_2 = ... = x_n = r$, connected in series, the value nr is obtained. For the same n resistors connected in parallel, the value r / n is obtained. Clearly, the value nr is exactly n^2 times larger than the value r / n. However, the prediction provided by inequality (3.1) is a much deeper result. *It is valid for any possible values for the resistances of the elements.* The comparison provided by inequality (3.1) does not require equal resistances.

3.1.2 Thermal Resistors and Electric Capacitors Connected in Series and Parallel

Identical reasoning applies to the problem related to the equivalent thermal resistance of n thermally conducting elements of different materials arranged in series (Figure 3.2a) and the equivalent thermal resistance of the n elements arranged in parallel (Figure 3.2b; Tipler and Mosca, 2008). The equivalent thermal resistance R_s of n elements arranged in series, with thermal resistances $R_1, R_2, \ldots, R_n$, is given by

$$R_s = R_1 + R_2 + \cdots + R_n \tag{3.3}$$

while the equivalent thermal resistance R_p of the same elements arranged in parallel is given by

$$R_p = \frac{1}{1/R_1 + 1/R_2 + \cdots + 1/R_n} \tag{3.4}$$

If x_i in inequality (3.1) stands for the thermal resistance of the ith element, inequality (3.1) now predicts a different physical property: the equivalent thermal resistance of n elements in series is at least n^2 times larger than the equivalent thermal resistance of the same elements arranged in parallel, irrespective of the individual thermal resistances of the elements.

Now consider n electric capacitors and suppose that x_i stands for the capacitance of the ith capacitor. The expression $x_1 + x_2 + \cdots + x_n$ on the left-hand side of inequality (3.1) can be interpreted as equivalent capacitance of n capacitors

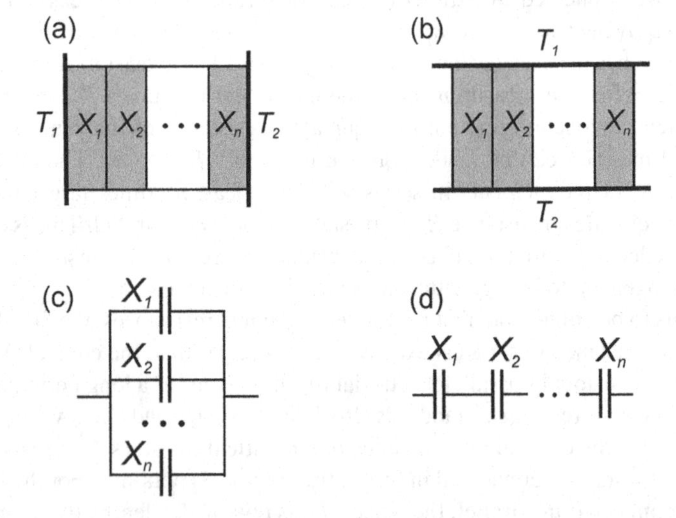

FIGURE 3.2 Interpretations of inequality 3.1 involving (a) elements arranged in series; (b) elements arranged in parallel; (c) capacitors arranged in parallel and (d) capacitors arranged in series

connected in parallel. On the right-hand side of the inequality, the expression $\frac{1}{1/x_1 + 1/x_2 + \cdots + 1/x_n}$ can be meaningfully interpreted as equivalent capacitance of the same n capacitors connected in series. Yet another physical property is predicted by inequality (3.1): the equivalent capacitance of n capacitors connected in parallel is at least n^2 times larger than the equivalent capacitance of the same capacitors connected in series, irrespective of the values of the individual capacitors.

The method of algebraic inequalities is domain-independent. It transcends mechanical engineering and can be used in many unrelated domains. In each application example, the left hand side of inequality (3.1) was interpreted as a property of a system with elements connected in series or parallel and the right-hand side of the inequality was interpreted as the same property characterising a system built on the same elements but connected in parallel or series configuration. The interpretation of inequality (3.1) establishes a link between the properties of the two different system configurations.

If multiple real experiments are performed in connecting elastic elements, resistors, thermal elements and capacitors in series and parallel and the equivalent stiffness, electrical resistance, thermal resistance and capacitance are measured, the observations will not contradict inequality (3.1). The statement expressed by inequality (3.1) and the behaviour of the systems of springs, resistors, thermal elements and capacitors arranged in series and parallel are consistent and no contradiction will be present between the prediction of the algebraic inequality and the behaviour of the physical systems.

These examples illustrate physical properties predicted from interpreting a correct algebraic inequality and the principle of non-contradiction.

3.2 CONSTRUCTING A SYSTEM WITH SUPERIOR RELIABILITY BY A MEANINGFUL INTERPRETATION OF ALGEBRAIC INEQUALITIES

3.2.1 Reliability of Systems with Components Logically Arranged in Series and Parallel

Before introducing interpretation of algebraic inequalities with the purpose of improving system reliability, the basics of evaluating the reliability of simple systems, with components logically arranged in series and parallel, will be covered first.

Consider a system including n independently working components. Let S denote the event 'the system is in working state at the end of a specified time interval' and C_k ($k = 1,2,\ldots,n$) denote the events 'component k is in working state at the end of the specified time interval'. For components logically arranged in series (Figure 3.3a), the system is in working state at the end of the specified time interval only if all components are in working state.

Reliability is the ability of an entity to work without failure for a specified time interval, under specified conditions and environment. The ability to work without

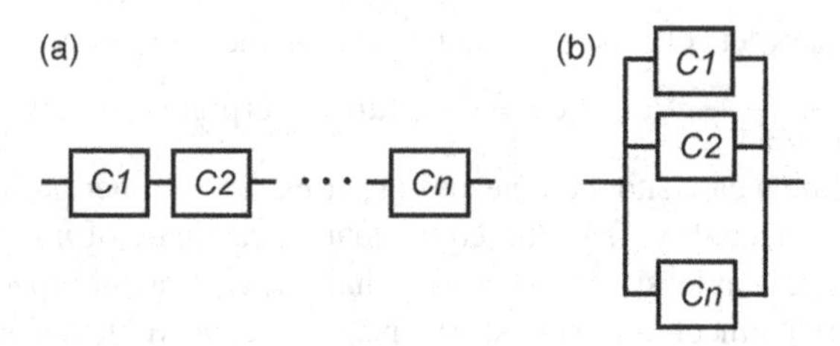

FIGURE 3.3 (a) A system with components (a) logically arranged in series and (b) logically arranged in parallel.

failure within a specified time interval is measured by the probability of working without failure during that interval.

According to the system reliability theory (Bazovsky, 1961; Hoyland and Rausand, 1994), the probability of system success (system in working state at the end of the specified time interval) is a product of the probabilities that the components will be in working state at the end of the specified time interval:

$$P(S) = P(C_1) \times P(C_2) \times \ldots \times P(C_n) \tag{3.5}$$

Denoting by R the probability $P(S)$ that the system will be in working state at the end of the specified time interval, and by $R_k = P(C_k)$ the probability that the kth component will be in working state at the end of the specified time interval, equation (3.5) becomes

$$R = R_1 \times R_2 \times \ldots \times R_n \tag{3.6}$$

In equation (3.6), R will be referred to as the reliability of the system and R_k as the reliability of the kth component.

Now consider independently working components logically arranged in parallel (Figure 3.3b). According to the system reliability theory (Bazovsky, 1961; Hoyland and Rausand, 1994), the probability of system success (system in working state) is equal to the probability that at least a single component will be in working state at the end of the specified time interval.

The event 'at least a single component will be in working state at the end of the specified time interval' and the event 'none of the components will be in working state at the end of that interval' are complementary events whose probabilities add up to unity. Therefore, the probability that at least a single component will be in working state can be evaluated by subtracting from unity the probability that none of the components will be in working state. The advantage offered by this inverse-thinking approach is that the probability that none of the components will be in working state at the end of the specified time interval is easy to calculate.

Indeed, if $R_1, R_2, \ldots, R_n$ denote the reliabilities of the separate components, the probability $P(\overline{S})$ that none of the components will be in working state at the end of the specified time interval (the probability of system failure) is given by

$$P(\bar{S}) = (1 - R_1)(1 - R_2)\ldots(1 - R_n) \tag{3.7}$$

Consequently, the probability $P(S)$ that the system will be in working state is given by

$$P(S) = 1 - P(\bar{S}) = R = 1 - (1 - R_1)(1 - R_2)\ldots(1 - R_n) \tag{3.8}$$

Note that for a logical arrangement of the components in series, the system reliability is a product of the reliabilities of the components, while for a logical arrangement of the components in parallel, the probability of system failure is a product of the probabilities of failure of the components.

A system with components logically arranged in series and parallel can be reduced in complexity in stages, as shown in Figure 3.4. In the first stage, the components in Figure 3.4a, logically arranged in series, with reliabilities R_1, R_2 and R_3 are reduced to an equivalent component with reliability $R_{123} = R_1 R_2 R_3$. The components logically arranged in series with reliabilities R_4 and R_5 are reduced to an equivalent component with reliability $R_{45} = R_4 R_5$ and the components from the parallel branch with reliabilities R_6 and R_7 are reduced to an equivalent component with reliability $R_{67} = R_6 R_7$. The resultant equivalent network is shown in Figure 3.4b.

In the second stage, the components in parallel with reliabilities R_{123} and R_{45} in Figure 3.4b are reduced to an equivalent component with reliability $R_{12345} = 1 - (1 - R_{123})(1 - R_{45})$. The resultant equivalent reliability network is shown in Figure 3.4c.

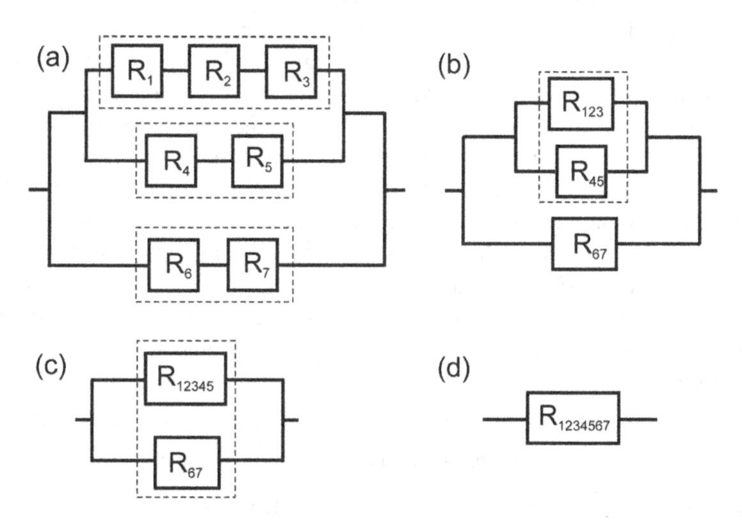

FIGURE 3.4 Network reduction method for determining the reliability of a system including components logically arranged in series and parallel (a) initial network; (b) the network after the first network reduction; (c) the network after the second network reduction; (d) the original network reduced to a single component.

Finally, the reliability network in Figure 3.4c is further simplified by reducing equivalent components with reliabilities R_{12345} and R_{67} to a single equivalent component with reliability $R_{1234567} = 1-(1-R_{12345})\times(1-R_{67})$ (Figure 3.4d). This is also the final result for the reliability of the original network in Figure 3.4a.

This is the essence of the *network reduction method* for determining the reliability of series–parallel systems where components are logically arranged only in series and parallel (Ebeling, 1997). Various techniques for determining system reliability of complex networks that are not with series–parallel topology have been considered in Todinov (2016).

3.2.2 Constructing a Series–Parallel System with Superior Reliability through Interpretation of an Algebraic Inequality

Inequalities involving terms of the type $1-a, 1-b, 1-c, 1-a^2, 1-b^2, 1-c^2, 1-ab, 1-ac, 1-bc$, where $0<a<1$, $0<b<1$ and $0<c<1$, can often be interpreted as reliabilities of series–parallel systems. The variables a, b and c can be interpreted as reliabilities of components from different types (e.g. type A, type B and type C).

As an example, consider the inequality

$$[1-(1-a)(1-c)]\times[1-(1-ab)(1-bc)] \geq [1-(1-a)(1-b)]\times[1-(1-ac)(1-bc)] \tag{3.9}$$

where $0<a<1$, $0<b<1$, $0<c<1$ and $a>b>c$. Both sides of inequality (3.9) can be interpreted as reliabilities of two series-parallel systems in two different configurations, built with the same components.

The left-hand side of this inequality can be interpreted as the reliability of the series–parallel system in Figure 3.5a and the right-hand side of the inequality can be interpreted as the reliability of the series–parallel system in Figure 3.5b.

Inequality (3.9) predicts that the reliability of the series–parallel system in Figure 3.5a is greater than the reliability of the system in Figure 3.5b.

Inequality (3.9), when tested with the simulation program in Chapter 2, for millions of combinations of random values for the reliabilities a, b and c, is never contradicted. This indicates that the inequality is probably true and the reliability of the system in Figure 3.5a is superior to the reliability of the system in Figure 3.5b, irrespective of the reliability values a, b and c of the components, as long as $a>b>c$ is fulfilled. After the confirmatory outcome from testing the inequality with the simulation program, an attempt for a rigorous proof can be made. Inequality (3.9) is equivalent to the inequality

$$(a+c-ac)\times(ab+bc-ab^2c) \geq (a+b-ab)(ac+bc-abc^2) \tag{3.10}$$

By factoring abc from both sides of inequality (3.10), it is not difficult to see that proving inequality (3.10) is equivalent to proving

$$[1+c/a-c]\times[a/c+1-ab] \geq [1+b/a-b]\times[a/b+1-ac]$$

which, in turn, is equivalent to proving the inequality

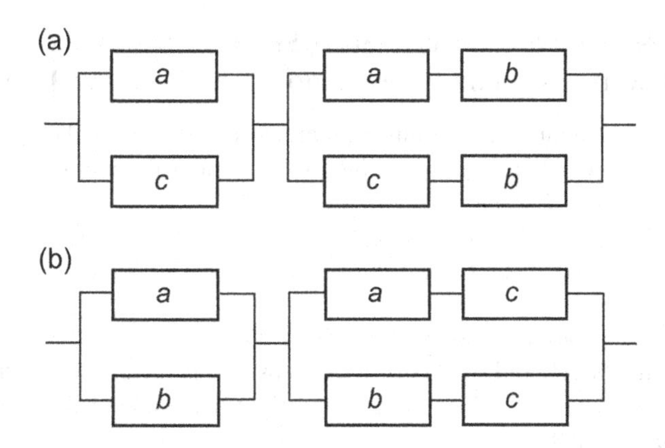

FIGURE 3.5 Competing series–parallel systems: (a) first system configuration versus (b) the second system configuration.

$$a/c - a/b - b/a + c/a + b + ac - ab - c \geq 0 \tag{3.11}$$

The left-hand side of inequality (3.11) can be presented as

$$a/c - a/b - b/a + c/a + b + ac - ab - c = (1-a)(b-c) + \frac{(b-c)(a^2 - bc)}{abc} \tag{3.12}$$

which is positive because both terms $(1-a)(b-c) \geq 0$ and $\frac{(b-c)(a^2 - bc)}{abc} \geq 0$ are positive (by definition, $a > b > c$).

This completes the rigorous proof that the reliability of the system in Figure 3.5a is superior to the reliability of the system in Figure 3.5b if $a>b>c$, irrespective of the actual reliability values a, b and c of the components building the systems.

The difference in the reliabilities of the two competing systems is significant and this can be seen after substituting the specific values $a = 0.8$, $b = 0.6$ and $c = 0.2$ for the reliabilities of components. For the left-hand side of inequality (3.9), the value

$$R_a = [1-(1-a)(1-c)] \times [1-(1-ab)(1-bc)] =$$

$$= [1-(1-0.8)(1-0.2)] \times [1-(1-0.8 \times 0.6)(1-0.6 \times 0.2)] = 0.456$$

is obtained for the reliability of the system in Figure 3.5a, while from the right-hand side of inequality (3.9), the value

$$R_b = [1-(1-a)(1-b)] \times [1-(1-ac)(1-bc)] =$$

$$= [1-(1-0.8)(1-0.6)] \times [1-(1-0.8 \times 0.2)(1-0.6 \times 0.2)] = 0.24$$

is obtained for the reliability of the system in Figure 3.5b.

3.2.3 Constructing a Parallel–Series System with Superior Reliability through Interpretation of an Algebraic Inequality

Often, additional manipulation must be applied in order to make the algebraic inequality interpretable as 'system reliability'. Consider the non-trivial algebraic inequality (Todinov, 2021)

$$(1-a^2b)(1-b^2c)(1-c^2a) \geq (1-a^3)(1-b^3)(1-c^3) \tag{3.13}$$

where a, b, c are positive numbers smaller than 1.

Multiplying the left and right part of (3.13) by '–1' reverses the direction of the inequality and by adding '1' to the left- and right-hand side of inequality (3.13), the equivalent inequality

$$1-(1-a^2b)(1-b^2c)(1-c^2a) \leq 1-(1-a^3)(1-b^3)(1-c^3) \tag{3.14}$$

is obtained.

A meaningful interpretation can be given to the left- and right-hand side of inequality (3.14) as reliabilities of two system configurations built with the same components.

Consider the system in Figure 3.6a built on three different types of components: A, B and C. The system from Figure 3.6a is in working state if there is path

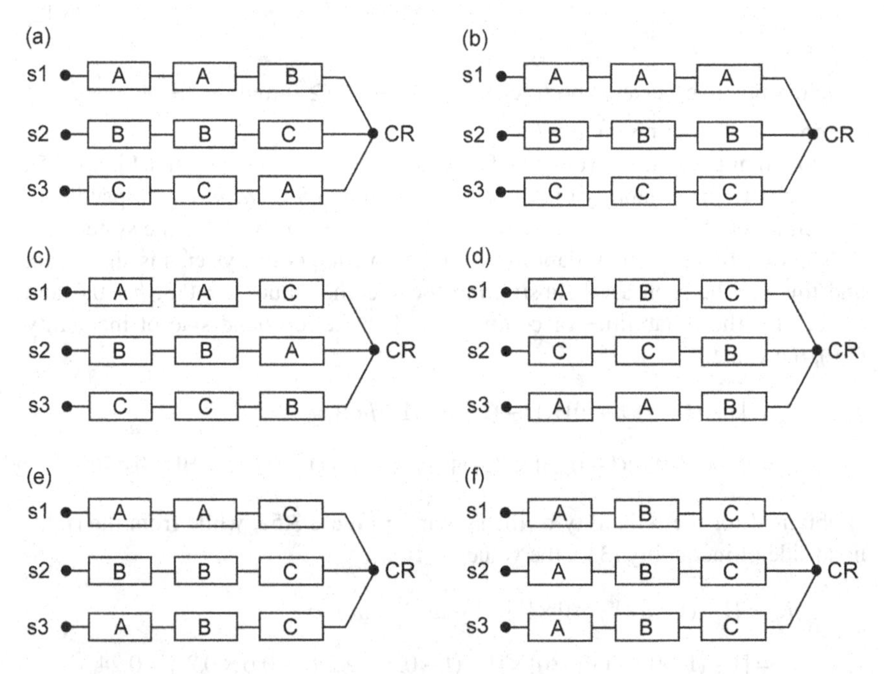

FIGURE 3.6 A parallel–series cooling system with different configuration of the sections: (a),(b),(c),(d),(e) and (f).

through working components from any of the source nodes $s1$, $s2$, $s3$ to the end node CR. In other words, all three components in any of the parallel lines must be in working state for the system to be in working state.

Systems with parallel–series arrangements of the type in Figure 3.6a are very common. Lines working in parallel and cooling a chemical reactor are examples of such systems. In a cooling line, the building sections are logically arranged in series (which means that a cooling line is working if all sections of the cooling line are in working state). The chemical reactor is adequately cooled if at least a single cooling line is in working state and delivers cooling fluid from any of the sources $s1$, $s2$, $s3$ to the chemical reactor CR. As a result, with respect to cooling the reactor CR, the cooling lines are arranged in parallel.

Now suppose that each cooling line contains three sections, each of which can be of different type: type A, type B and type C. Let a, b and c stand for the reliabilities of components of types A, B and C in the system configurations from Figure 3.6. The left-hand side of inequality (3.14) gives the reliability of the system configuration in Figure 3.6a, while the right-hand side of inequality (3.14) gives the reliability of the system configuration in Figure 3.6b. Inequality (3.14) predicts that the reliability of the system configuration in Figure 3.6b is greater than the reliability of the system configuration in Figure 3.6a.

Inequality (3.14), when tested with the simulation program from Chapter 2, for millions of combinations of random values for the reliabilities a, b and c, is never contradicted. This indicates that the inequality is probably true and the reliability of the system in Figure 3.6b is, indeed, greater than the reliability of the system in Figure 3.6a, irrespective of the reliability values a, b and c of the components. After the confirmatory outcome from testing inequality (3.14), an attempt can be made for a rigorous proof.

Proving inequality (3.14) is equivalent to proving inequality (3.13). This inequality can be proved by combining the methods of symmetry and the AM–GM inequality (Todinov, 2021).

Using the symmetry of the inequality, without loss of generality, $a \geq b \geq c \geq 0$ can be assumed. Inequality (3.13) can be proved if the equivalent inequality

$$\sqrt[3]{\frac{1-a^3}{1-a^2b} \times \frac{1-b^3}{1-b^2c} \times \frac{1-c^3}{1-c^2a}} \leq 1 \tag{3.15}$$

can be proved.

From the AM–GM inequality, it follows that

$$\sqrt[3]{\frac{1-a^3}{1-a^2b} \times \frac{1-b^3}{1-b^2c} \times \frac{1-c^3}{1-c^2a}} \leq \frac{1}{3}\left(\frac{1-a^3}{1-a^2b} + \frac{1-b^3}{1-b^2c} + \frac{1-c^3}{1-c^2a}\right)$$

Consequently, if inequality (3.16)

$$\frac{1}{3}\left(\frac{1-a^3}{1-a^2b} + \frac{1-b^3}{1-b^2c} + \frac{1-c^3}{1-c^2a}\right) \leq 1 \tag{3.16}$$

can be proved, inequalities (3.15) and (3.14) will follow.

Proving inequality (3.16) is reduced to proving the equivalent inequality

$$\frac{1-a^3}{1-a^2b}+\frac{1-b^3}{1-b^2c}+\frac{1-c^3}{1-c^2a}\le 1+1+1$$

which is equivalent to the inequality

$$\frac{a^2(a-b)}{1-a^2b}+\frac{b^2(b-c)}{1-b^2c}+\frac{c^2(c-a)}{1-c^2a}\ge 0 \tag{3.17}$$

To prove inequality (3.17), the techniques 'conditioning on mutually exclusive cases' and 'proving an intermediate inequality' will be applied.

Consider two mutually exclusive cases, related to the variables a, b and c: (i) $b^2 \le ac$ and (ii) $b^2 > ac$.

Let $b^2 \le ac$ hold (case i). In this case, $b^2c \le ac^2$ also holds and the left-hand side of (3.17) will be decreased if the denominator $1-a^2b$ in the first term of (3.17) is replaced by the larger value $1-b^2c$, the denominator $1-c^2a$ of the third term is replaced by the larger value $1-b^2c$ and b^2 in the numerator of the second term is replaced by the smaller factor c^2 ($c<b$). As a result,

$$\frac{a^2(a-b)}{1-a^2b}+\frac{b^2(b-c)}{1-b^2c}+\frac{c^2(c-a)}{1-c^2a}\ge\frac{a^2(a-b)}{1-b^2c}+\frac{c^2(b-c)}{1-b^2c}+\frac{c^2(c-a)}{1-b^2c} \tag{3.18}$$

holds. Factoring c^2 from the second- and third term of the right-hand side of inequality (3.18) results in:

$$\frac{a^2(a-b)}{1-b^2c}+\frac{c^2(b-c)}{1-b^2c}+\frac{c^2(c-a)}{1-b^2c}=\frac{a^2(a-b)}{1-b^2c}+\frac{c^2(b-a)}{1-b^2c}=(a-b)\frac{a^2-c^2}{1-b^2c} \tag{3.19}$$

for the right-hand side.

Since $a^2-c^2\ge 0$, $a-b\ge 0$ and $1-b^2c\ge 0$, the right-hand side of (3.19) is greater than 0 or equal to 0. The intermediate inequality $(a-b)\frac{a^2-c^2}{1-b^2c}\ge 0$ is true, which shows that if case (i) holds, inequality (3.17), the equivalent inequalities (3.16), (3.15) and the original inequality (3.14) also hold.

Now let $b^2 > ac$ hold (case ii). In this case, $b^2c > ac^2$ holds and the left-hand side of (3.17) will be decreased if the denominator $1-a^2b$ in the first term is replaced by the larger value $1-c^2a$, the denominator $1-b^2c$ of the second term is replaced by the larger value $1-c^2a$ and b^2 in the numerator of the second term is replaced by c^2. As a result,

$$\frac{a^2(a-b)}{1-a^2b}+\frac{b^2(b-c)}{1-b^2c}+\frac{c^2(c-a)}{1-c^2a}>\frac{a^2(a-b)}{1-c^2a}+\frac{c^2(b-c)}{1-c^2a}+\frac{c^2(c-a)}{1-c^2a} \tag{3.20}$$

holds. Factoring c^2 from the second- and third term of the right-hand side of inequality (3.20) results in:

$$\frac{a^2(a-b)}{1-c^2a}+\frac{c^2(b-c)}{1-c^2a}+\frac{c^2(c-a)}{1-c^2a}=\frac{a^2(a-b)}{1-c^2a}+\frac{c^2(b-a)}{1-c^2a}=(a-b)\frac{a^2-c^2}{1-c^2a} \tag{3.21}$$

Since $a^2-c^2\geq 0$, $a-b\geq 0$ and $1-c^2a\geq 0$, the right part of (3.21) is non-negative. This proves the intermediate inequality $(a-b)\frac{a^2-c^2}{1-c^2a}\geq 0$, which shows that if case (ii) holds, inequality (3.17) and the equivalent inequalities (3.15) and the original inequality (3.14) also hold. The proofs of the two mutually exclusive and exhaustive cases prove inequality (3.14). The system configuration in Figure 3.6b is intrinsically more reliable than the system configuration in Figure 3.6a.

The difference in the reliabilities of the two competing system configurations is significant and this can be seen after substituting specific values ($a = 0.8$, $b = 0.6$ and $c = 0.2$) for the reliabilities of components. From the right-hand side of inequality (3.14), the value

$$R_b = 1-(1-a^3)(1-b^3)(1-c^3) = 0.62$$

is obtained for the reliability of the system configuration in Figure 3.6b, while from the left-hand side of inequality (3.14), the value

$$R_a = 1-(1-a^2b)(1-b^2c)(1-c^2a) = 0.45$$

is obtained for the reliability of the system configuration in Figure 3.6a.

In a similar fashion, it can be shown that the inequalities

$$1-(1-a^2c)(1-b^2a)(1-c^2b)\leq 1-(1-a^3)(1-b^3)(1-c^3) \tag{3.22}$$

$$1-(1-abc)(1-c^2b)(1-a^2b)\leq 1-(1-a^3)(1-b^3)(1-c^3) \tag{3.23}$$

$$1-(1-abc)(1-a^2c)(1-b^2c)\leq 1-(1-a^3)(1-b^3)(1-c^3) \tag{3.24}$$

$$1-(1-abc)(1-abc)(1-abc)\leq 1-(1-a^3)(1-b^3)(1-c^3) \tag{3.25}$$

are also true.

The left-hand side of inequalities (3.22)–(3.25) represents the reliability of the system configurations in Figure 3.6c, d, e, f, correspondingly, while the right-hand side of inequalities (3.22)–(3.25) represents the reliability of the system configuration in Figure 3.6b.

As a result, a general conclusion can be made that the parallel–series system configuration in Figure 3.6b is intrinsically more reliable than the system configurations in Figures 3.6a, 3.6c, 3.6d, 3.6e and 3.6f composed of the same components because the direction of the inequalities is preserved for any combination of values for the reliabilities a, b and c.

For a cooling system composed of three new sections A, three medium-age sections B and three old sections C, arranging all new sections A in the same cooling branch, all medium sections B in another cooling branch and all old sections C in a separate cooling branch (Figure 3.6b) maximises the reliability of the system.

3.3 SELECTING THE SYSTEM WITH SUPERIOR RELIABILITY THROUGH INTERPRETATION OF THE INEQUALITY OF NEGATIVELY CORRELATED EVENTS

This interpretation technique involves the inequality of negatively correlated events, introduced in Todinov (2019c).

Consider m independent events A_1, A_2,..., A_m that are not mutually exclusive. This means that there are at least two events A_i and A_j for which $P(A_i \cap A_j) \neq \varnothing$. It is known with certainty that if any particular event A_k from the set of events does not occur ($k = 1,\dots,m$), then at least one of the other events occurs. In other words, the relationship

$$P(A_1 \cup \dots \cup \bar{A}_k \cup \dots \cup A_m) = 1 \tag{3.26}$$

holds for the set of m events.

Under these assumptions, it can be shown that the next inequality holds:

$$P(A_1) + P(A_2) + \dots + P(A_m) > 1 \tag{3.27}$$

where $P(A_i)$, $i = 1,...,m$, stands for the probability of event A_i.

Inequality (3.27) will be referred to as *the inequality of negatively correlated events.*

To prove this inequality, consider the number of outcomes n_1, n_2, …, n_m leading to the separate events A_1, A_2, …, A_m, correspondingly. Let n denote the total number of possible outcomes. Note that according to the definition of negatively correlated events, at least one of the events must occur. From the definition of inversely correlated events, it follows that any of the n distinct outcomes related to the state of the m events (occurrence/non-occurrence) corresponds to the occurrence of at least one event A_i. Since at least two events A_i and A_j can occur simultaneously, the sum of the outcomes leading to the separate events A_1, A_2, …, A_m is greater than the total number of outcomes n:

$$n_1 + n_2 + \dots + n_m > n \tag{3.28}$$

This is because of the condition that at least two events A_i and A_j can occur simultaneously. Then, at least one outcome must be counted twice: once for event A_i and once for event A_j. Dividing both sides of inequality (3.28) by the positive value n does not alter the direction of inequality (3.28), and the result is the inequality

$$n_1/n + n_2/n + \cdots + n_m/n > 1 \tag{3.29}$$

which is inequality (3.27).

The power of the simple inequality (3.27) will be demonstrated by interpreting it for two events only (A_1 and A_2). Event A_1 stands for 'system 1 is working at the end of a specified time interval', while event A_2 stands for 'system 2 is not working at the end of the specified time interval'. In addition, it is known that if system 1 is not working, system 2 is not working and also if system 2 is working, system 1 is also working.

The usefulness of such interpretation of inequality (3.27) will be demonstrated on two competing systems, including the same components, whose reliability networks are shown in Figure 3.7.

Despite the deep uncertainty related to the reliability of the components building the systems, the reliabilities of the systems can still be ranked, by a meaningful interpretation of the inequality of negatively correlated events.

Let event $A_1 \equiv A$ stand for 'the system in Figure 3.7a is working at the end of a specified time interval', while event $A_2 \equiv \overline{B}$ stand for 'the system in Figure 3.7b is not working at the end of the specified time interval' ($P(B) + P(\overline{B}) = 1$) (Figure 3.7).

The conditions of inequality (3.27) are fulfilled for events A and $\overline{B}$ related to the systems in Figure 3.7.

Indeed, if event $\overline{B}$ does not occur, this means that system (3.7b) is working (event B occurs). This can happen only if there is a path through working components from the start node s to the end node t. The existence of a path through working components for the system in Figure 3.7b from the start node s to the end node t means that either components 1,2,5 are working or components 1,3,4 or components 1,6 are working.

Components 1,2,5 working for the system in Figure 3.7b implies that the system in Figure 3.7a is working because there will be the path 1,5 through working components connecting the start node s and the end node t in Figure 3.7a. Similarly, components 1,3,4 working in the system in Figure 3.7b implies that the system

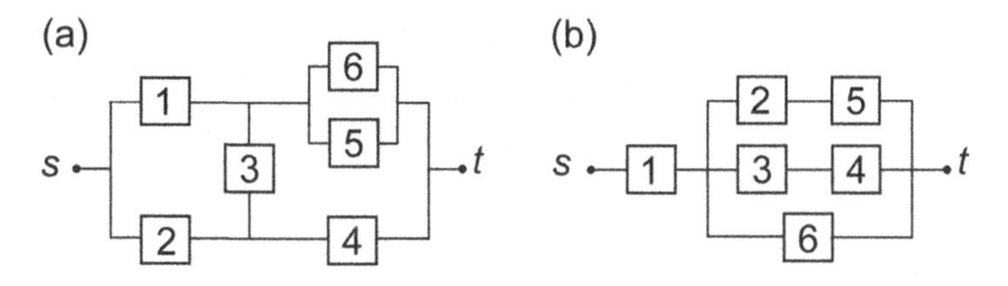

FIGURE 3.7 Ranking the reliabilities of two systems with unknown reliabilities of their components, (a) first alternative versus (b) second alternative.

in Figure 3.7a is working because there will be the path 1,3,4 through working components in Figure 3.7a connecting the start node s and the end node t. Finally, components 1,6 in Figure 3.7b working implies that the system in Figure 3.7a is working because there will be the path 1,6 connecting the start node s and the end node t. As a result, if event $\bar{B}$ does not occur, then event A occurs.

Conversely, if event A does not occur, then there will be no path through working components from the start node s to the end node t in Figure 3.7a. The absence of a path through working components implies that the components from at least one of the following sets: (1,2), (1,3,4), (2,3,6,5), (6,5,4) are all in failed state. It can be seen that the existence of any of these sets of failed components implies the absence of a path through working components from the start node s to the end node t in Figure 3.7b.

This means that if the system in Figure 3.7a does not work, the system (3.7b) does not work either (the event $\bar{B}$ occurs). At the same time, both events can occur simultaneously ($P(A \cap \bar{B}) \neq 0$). This is, for example, the case if components 2,4 are in working state at the end of the specified time interval and component 1 is in a failed state.

The conditions of inequality (3.27) are fulfilled, therefore,

$$P(A) + P(\bar{B}) > 1 \tag{3.30}$$

holds, which is equivalent to

$$P(A) > 1 - P(\bar{B}) = P(B)$$

This means that $P(A) > P(B)$ irrespective of the reliabilities $r_1, r_2, r_3, r_4, r_5, r_6$ of components (1–6) building the systems. The meaningful interpretation of the inequality of negatively correlated events helped reveal the intrinsic reliability of competing systems and rank these in terms of reliability, in the absence of knowledge related to the reliabilities of their components.

4 Enhancing Systems Performance by Interpretation of the Bergström Inequality

4.1 EXTENSIVE QUANTITIES AND ADDITIVITY

Extensive quantities are typical examples of additive quantities. Extensive quantities change with changing the size of their supporting objects/systems (DeVoe, 2012; Mannaerts, 2014) and can be found in all areas of science and engineering. DeVoe (2012), for example, introduced 'extensivity' test that consists of dividing the system by an imaginary surface into two parts. Any quantity characterising the system that is the sum of the same quantity characterising the two parts is an extensive quantity and any quantity that has the same value in each part of the system is an intensive quantity.

Extensivity always implies additivity. Extensive quantities are, for example, mass, weight, amount of substance, number of particles, volume, distance, energy (kinetic energy, gravitational energy, electric energy, elastic energy, surface energy, internal energy), work, power, heat, force, momentum, electric charge, electric current, heat capacity, electric capacity, resistance (when the elements are in series), enthalpy and fluid flow. These are also examples of additive quantities.

Intensive quantities characterise the object/system locally and do not change with changing the size of the supporting objects/systems. Additivity is not present for intensive quantities. Additivity, for example, is not present for pressure or temperature. Consider, for example, a pressure vessel containing gas at a particular pressure and temperature. If a notional division of the pressure vessel is made, the temperature or pressure measured in the pressure vessel is not a sum of the temperature/pressure measured in the different parts of the vessel. Other properties where additivity is not present are 'density', 'concentration', 'hardness', 'velocity', 'acceleration', 'stress', 'surface tension', etc.

It is important to note that proportionality to mass is not a necessary condition for additivity. For a large group of extensive quantities (area, work, electric energy, elastic energy, displacement energy, surface energy, electric current, power, heat, electric charge), the proportionality to mass is absent (Mannaerts, 2014).

DOI: 10.1201/9781003199830-4

Depending on how the different elements composing a system are arranged, additivity may be present or absent. Thus, for resistances connected in series, additivity is present because the equivalent resistance of elements connected in series is a sum of the individual resistances. For resistances connected in parallel, additivity is absent.

Similarly, for voltage sources connected in series, additivity is present because the total voltage is a sum of the voltages of the individual sources. Additivity is also present for capacitors connected in parallel. For such an arrangement, the equivalent capacitance of the assembly is a sum of the individual capacitances of the capacitors. For capacitors connected arranged in series, additivity is absent.

Normally, dividing two extensive quantities gives an *intensive* quantity (such as obtaining density by dividing mass to volume) and, as a result, additivity is absent but this is not always the case. Thus, dividing the extensive quantity 'volume' by the extensive quantity 'area' gives the extensive quantity length. Additionally, dividing the extensive quantity 'work' by the extensive quantity 'distance' over which the work has been done gives the extensive quantity 'force'.

4.2 MEANINGFUL INTERPRETATION OF THE BERGSTRÖM INEQUALITY TO MAXIMISE ELECTRIC POWER OUTPUT

A special case of the general sub-additive function (2.38) is the Bergström inequality (Sedrakyan & Sedrakyan, 2010; Pop, 2009)

$$\frac{(a_1+a_2+\cdots+a_n)^2}{b_1+b_2+\cdots+b_n} \le \frac{a_1^2}{b_1}+\frac{a_2^2}{b_2}+\cdots+\frac{a_n^2}{b_n} \tag{4.1}$$

valid for any sequences $a_1,a_2,\ldots,a_n$ and $b_1,b_2,\ldots,b_n$ of positive real numbers. The role of the sub-additive function $f(a,b)$ in inequality (2.38) is played by the function $f(a,b)\equiv a^2/b$ in inequality (4.1).

The Bergström inequality (4.1) is a transformation of the well-known Cauchy–Schwarz inequality (Steele, 2004) stating that for any two sequences of real numbers $x_1,x_2,\ldots,x_n$ and $y_1,y_2,\ldots,y_n$, the following inequality holds:

$$(x_1y_1+x_2y_2+\cdots+x_ny_n)^2 \le (x_1^2+x_2^2+\cdots+x_n^2)(y_1^2+y_2^2+\cdots+y_n^2) \tag{4.2}$$

Indeed, if the substitutions $x_i=\frac{a_i}{\sqrt{b_i}}$ $(i=1,\ldots,n)$ and $y_i=\sqrt{b_i}$ $(i=1,\ldots,n)$ are made in the Cauchy–Schwarz inequality (4.2), the result is the inequality

$$\left(\frac{a_1}{\sqrt{b_1}}\sqrt{b_1}+\frac{a_2}{\sqrt{b_2}}\sqrt{b_2}+\cdots+\frac{a_n}{\sqrt{b_n}}\sqrt{b_n}\right)^2 \le \left((a_1/\sqrt{b_1})^2+\cdots+(a_n/\sqrt{b_n})^2\right)$$

$$\left((\sqrt{b_1})^2+\cdots+(\sqrt{b_n})^2\right)$$

which leads to inequality (4.1), valid for any sequences $a_1, a_2, \ldots, a_n$ and $b_1, b_2, \ldots, b_n$ of positive real numbers. Equality in inequality (4.1) is attained only if $a_1 / b_1 = a_2 / b_2 = \ldots = a_n / b_n$.

Inequality (4.1) has a number of interesting potential applications related to increasing the effect of additive quantities (factors).

Inequality (4.1) effectively states that the effect of the additive quantities $a = \sum_{i=1}^{n} a_i$ and $b = \sum_{i=1}^{n} b_i$ can be increased by segmenting them into smaller parts a_i and b_i, $i = 1, \ldots, n$ and accumulating their individual effects a_i^2 / b_i, represented by the terms on the right-hand side of inequality (4.1).

Inequality (4.1) has a universal application in science and technology as long as the variables a_i, b_i and the terms a_i^2 / b_i are additive quantities and have meaningful interpretation. To extract new knowledge from the interpretation of inequality (4.1), there is no need of any forward analysis. The condition for applying inequality (4.1) is the possibility to present an additive quantity p_i as a ratio of a square of an additive quantity a_i and another additive quantity b_i:

$$p_i = a_i^2 / b_i \tag{4.3}$$

As an example, consider a case where factor a is 'voltage' from a source whose elements are arranged in series (an additive quantity) and factor b is 'resistance' of elements arranged in series (also an additive quantity). Suppose that the source of voltage V has been applied to n elements in series, with resistances $r_1, r_2, \ldots, r_n$ (Figure 4.1a).

Consider segmenting the source of voltage V into n smaller sources with voltages v_i such that $V = v_1 + \cdots + v_n$ (Figure 4.1b). Let $a_i = v_i$, $i = 1, \ldots, n$ be the smaller voltages applied to n separate elements with resistances r_i (Figure 4.1b). If $b_i = r_i$, condition (4.3) is fulfilled because the equation

$$p_i = v_i^2 / r_i \tag{4.4}$$

holds, where p_i is the power output [W] created by the ith voltage source, v_i is the voltage of the ith voltage source and r_i is the resistance of the ith element. In equation (4.4), all variables v_i, p_i and r_i are additive quantities.

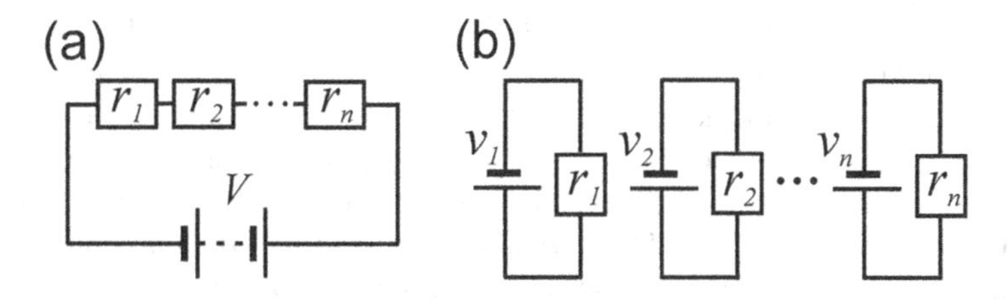

FIGURE 4.1 (a) A design option consisting of a single voltage source applied to elements connected in series; (b) a competing design option consisting of a voltage source V segmented into n smaller sources v_i applied to the individual elements.

Inequality (4.1) can be rewritten as

$$\frac{V^2}{r_1 + r_2 + \ldots + r_n} \leq \frac{v_1^2}{r_1} + \frac{v_2^2}{r_2} + \ldots + \frac{v_n^2}{r_n} \tag{4.5}$$

The left-hand side of inequality (4.5) can now be interpreted as power dissipated in the circuit from Figure 4.1a and the right-hand part can be interpreted as total power dissipated in the individual circuits from Figure 4.1b (Todinov, 2020d). The prediction from inequality (4.5) is: *The power output from a source with voltage V on elements connected in series is smaller than the total power output from the sources with voltages* $v_i \left(\sum_i v_i = V \right)$, *resulting from segmenting the original source V and applying the segmented sources to the individual elements.*

This prediction holds irrespective of the individual resistances r_i of the elements and the voltage sources v_i into which the original voltage V has been segmented.

Note that the existence of asymmetry in the system is essential for increasing the power output through segmentation of the voltage source and the resistances. No increase in the power output is present if

$$v_1 / r_1 = v_2 / r_2 = \ldots = v_n / r_n = i \tag{4.6}$$

is fulfilled.

This means that combinations of resistances and sources resulting in the same current i in the segmented circuits do not bring an increase of the total power output.

To maximise the right-hand side of inequality (4.5), the squared voltages arranged in descending order $v_1^2 \geq v_2^2 \geq \ldots \geq v_n^2$ must be paired with the resistances arranged in ascending order ($r_1 \leq r_2 \leq \ldots \leq r_n$). For resistances arranged in ascending order, their reciprocals are arranged in descending order ($1/r_1 \geq 1/r_2 \geq \ldots \geq 1/r_n$), and according to the rearrangement inequality (see Section 2.1.5), the dot product $v_1^2(1/r_1) + v_2^2(1/r_2) + \cdots + v_n^2(1/r_n)$ of two similarly ordered sequences is maximal.

The effect of voltage and resistance segmentation is significant. Thus, for resistances $r_1 = 10\Omega, r_2 = 15\Omega, r_3 = 25\Omega, r_4 = 50\Omega$ and voltage source of 16 V, segmented into sources with voltages $v_1 = 6\text{V}$, $v_2 = 5\text{V}$, $v_3 = 3\text{V}$ and $v_4 = 2\text{V}$, the maximum possible power output is $P_{\max} = v_1^2 / r_1 + v_2^2 / r_2 + v_3^2 / r_3 + v_4^2 / r_4 = 5.7\,\text{W}$. Any other permutation of voltages and resistances results in a smaller power output.

If the voltage source $V = 16\text{V}$ is not segmented, applying the voltage V to the four elements connected in series delivers output power $P = V^2 / (r_1 + r_2 + r_3 + r_4) = 2.56\,\text{W}$, which is less than half the maximum power of $P_{\max} = 5.7\,\text{W}$ delivered in the case of a voltage source segmentation.

The result from the meaningful interpretation of the algebraic inequality can be applied in electric circuits for heating. Despite that the circuits in Figure 4.1a and b seem to be very different electric systems, they both can be viewed as alternative design options of a heating system. The circuit in Figure 4.1b dissipates more power (heat) for the same total number of voltage elements and the same set of resistive elements. It needs to be pointed out again that this advantage is not present if the segmented circuits are characterised by the same currents through the resistive elements.

4.3 MEANINGFUL INTERPRETATION OF THE BERGSTRÖM INEQUALITY TO MAXIMISE THE STORED ELECTRIC ENERGY IN CAPACITORS

Another relevant meaning can be created for the variables a_i, b_i in inequality (4.1) if variable a stands for 'electric charge' and variable b stands for 'capacitance'.

In basic electronics (Floyd and Buchla, 2014), a well-known result is that the potential energy U stored by a charge Q in a capacitor with capacitance C is given by $U = Q^2 / (2C)$. Note that the potential energy U, the charge Q and capacitance C of elements arranged in parallel are all additive quantities.

Consider segmenting the charge Q into n smaller charges q_i, where i=1,..., n, such that $Q = q_1 + \cdots + q_n$. Let $a_i = q_i$, where $i = 1,\ldots,n$, in inequality (4.1) be the charges applied to n separate capacitors with capacitances C_i whose sum is equal to the capacitance of the original capacitor: $C = C_1 + C_2 + \cdots + C_n$. If $b_i = C_i$, inequality (4.1) can be rewritten as

$$\frac{Q^2}{2(C_1 + C_2 + \cdots + C_n)} \le \frac{q_1^2}{2C_1} + \frac{q_2^2}{2C_2} + \cdots + \frac{q_n^2}{2C_n} \tag{4.7}$$

Because $q_i^2 / (2C_i)$, q_i, C_i are all additive quantities, both sides of inequality (4.7) have meaningful interpretation. The left-hand side of the inequality can be interpreted as the energy stored by a charge Q in a capacitor with capacitance C, while the right-hand side can be interpreted as the total energy stored in multiple capacitors with capacitances C_i, $\sum_i C_i = C$, by smaller charges q_i, $\sum_{i=1}^{n} q_i = Q$, and applying these to the individual capacitors (Todinov, 2020d).

Inequality (4.1) now predicts that *the energy stored by a charge Q in a capacitor with capacitance C is smaller than the total energy stored in multiple capacitors with capacitances C_i, $\sum_i C_i = C$, by segmenting the charge Q into smaller charges and applying these to the individual capacitors.*

This prediction holds irrespective of the individual capacitances C_i of the capacitors and the charges q_i into which the initial charge Q has been segmented.

Note that existence of asymmetry is absolutely essential for increasing the electrical energy stored by segmenting the initial charge Q. No increase in the stored electrical energy is present if $q_1 / C_1 = q_2 / C_2 = \ldots = q_n / C_n = v$. This means that capacitors loaded to the same potential difference $v = q_i / C_i$ on the plates do not yield an increase of the amount of stored total electrical energy.

To maximise the right-hand side of inequality (4.7), the squared segmented charges arranged in descending order $q_1^2 \geq q_2^2 \geq \ldots \geq q_n^2$ must be paired with the capacitances arranged in ascending order ($C_1 \leq C_2 \leq \ldots \leq C_n$). For capacitances arranged in ascending order, the reciprocals $1/(2C_1) \geq 1/(2C_2) \geq \ldots \geq 1/(2C_n)$ are arranged in descending order and, according to the rearrangement inequality, the 'dot product' $q_1^2/(2C_1) + q_2^2/(2C_2) + \cdots + q_n^2/(2C_n)$ of two similarly ordered sequences is maximal (see Section 2.1.5).

The segmentation of additive quantities through the algebraic inequality (4.1) can be used to achieve systems and processes with superior performance and the algebraic inequality is applicable in any area of science and technology, as long as the additive quantities comply with the simple condition (4.3).

It is important to note that inequality (4.1) can be applied to each of the individual terms a_i^2/b_i on the left-hand side which, in turn, can be segmented. The result from this recursive segmentation is further multiplication of the effect from the segmentation.

Interestingly, no such properties have been reported in modern comprehensive texts in the field of electronics (Floyd and Buchla, 2014; Horowitz and Hill, 2015). This demonstrates that the meaningful interpretation of the abstract inequality (4.1) helped find an overlooked property in such a mature field like electronics.

4.4 AGGREGATION OF THE APPLIED VOLTAGE TO MAXIMISE THE ENERGY STORED IN CAPACITORS

It may appear that the segmentation of the applied voltage will maximise the energy stored in a set of capacitors similar to the way a segmentation of the applied voltage maximises the power on a set of resistors. Interestingly, this conjecture is false.

Indeed, the electrical energy U stored in a set of capacitors with capacitances $C_1, C_2, \ldots, C_n$, arranged in parallel (Figure 4.2) by an applied voltage of magnitude V, is given by (Floyd and Buchla, 2014)

$$U_1 = \frac{1}{2} C V^2 \tag{4.8}$$

where $C = C_1 + C_2 + \cdots + C_n$.

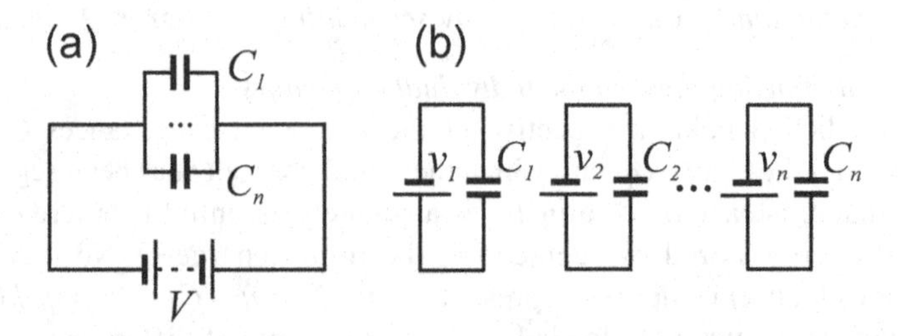

FIGURE 4.2 (a) A single voltage source applied to capacitors connected in parallel; (b) a voltage source V segmented into n smaller sources v_i applied to the individual capacitors.

Segmenting the voltage source V and applying the smaller voltage sources v_i to the individual capacitors result in total stored energy U_2 given by

$$U_2 = \frac{1}{2}C_1 v_1^2 + \frac{1}{2}C_2 v_2^2 + \cdots + \frac{1}{2}C_n v_n^2 \tag{4.9}$$

By a direct expansion, it can be seen that

$$\frac{1}{2}C_1 v_1^2 + \frac{1}{2}C_2 v_2^2 + \cdots + \frac{1}{2}C_n v_n^2 \leq \frac{1}{2}(C_1 + C_2 + \cdots C_n)(v_1 + v_2 + \cdots + v_n)^2 \tag{4.10}$$

Therefore, $U_2 \leq U_1$ which means that the segmentation of a voltage source over a number of capacitors is associated with a decrease of the total stored electrical energy. In the case of voltage sources applied to capacitors, the aggregation of the voltage sources and the capacitors leads to an increase of the total stored energy. Inequality (4.10) is based on the multivariable super-additive function $f(C,v) = (1/2)C v^2$

4.5 MEANINGFUL INTERPRETATION OF THE BERGSTRÖM INEQUALITY TO INCREASE THE ACCUMULATED ELASTIC STRAIN ENERGY

4.5.1 Increasing the Accumulated Elastic Strain Energy for Components Loaded in Tension

An alternative meaning can be created for the sub-additive function (4.1) if variable a, for example, stands for the additive quantity 'force' and variable b stands for the additive quantity 'area'.

It is a well-known result from mechanics of materials (Hearn, 1985) that the accumulated elastic strain energy U of a linearly elastic bar with length L and cross-sectional area A is given by the equation:

$$U = \frac{P^2 L}{2EA} \tag{4.11}$$

where E [Pa] denotes the Young's modulus of the material and P [N] is the magnitude of the loading force. The strain energy U [J] is an additive quantity.

Equation (4.11) can be written as

$$U = \frac{P^2}{2k} \tag{4.12}$$

where $k = EA / L$ [N/m] is the stiffness of the bar.

Consider the two system configurations in Figure 4.3. In the system configuration in Figure 4.3a, a single force P acts on a single large bar with cross-sectional area A. In the system configuration in Figure 4.3b, the original bar has been

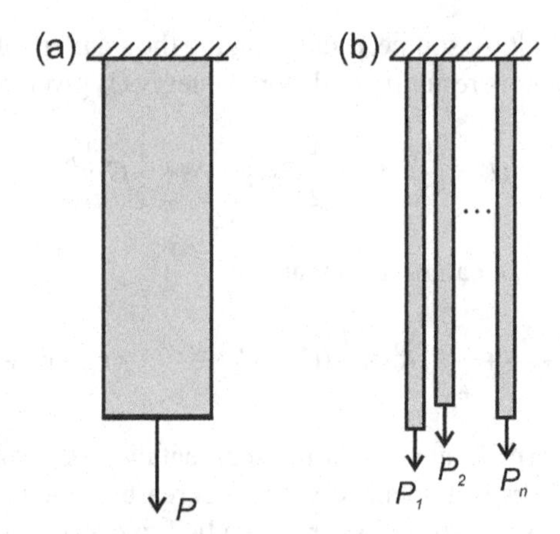

FIGURE 4.3 (a) A force P applied to a single bar; (b) segmented forces P_i applied to the individual bars into which the original bar has been segmented (the individual bars are loaded independently and they do not necessarily have the same displacements and stresses).

segmented into n individual bars with smaller cross sections $A_1, A_2, \ldots, A_n$, the sum of which is equal to the cross section A of the original bar ($A = A_1 + \cdots + A_n$) and the force P has also been segmented into n forces $P_1, P_2, \ldots, P_n$, the sum of which is equal to the initial loading force P ($P = P_1 + P_2 + \cdots + P_n$). The smaller forces P_i have been applied to the individual bars independently, and the individual bars do not necessarily have the same displacements and stresses (Figure 4.3b). The support is rigid and does not deform due to the segmented forces.

A question of interest is which system configuration in Figure 4.3 is capable of accumulating more elastic strain energy.

Let $a_i = P_i, i = 1, \ldots, n$ be the forces applied to the centroids of the n separate bars whose stiffness values are $k_i = E_i A_i / L_i$ (Figure 4.3b). Let also $b_i = k_i, i = 1, \ldots, n$.

The system configurations in Figure 4.3a and b are not equivalent mechanically and the equivalence is not the purpose of the load segmentation. The purpose of the load segmentation and the section segmentation is to increase the elastic strain energy stored in the system. This is why there is no requirement about equivalence of moments. The only requirement is the sum of the magnitudes of the segmented forces P_i to be equal to the magnitude of the original loading force P and the sum of the cross-sectional areas A_i of the segmented bars to be equal to the cross-sectional area A of the original bar.

Because the supports in Figure 4.3a and b are rigid, no deformations exist in the supports and the elastic strain energy of the system of bars is a sum of the elastic strain energies $P_i^2 / (2k_i)$ of the individual bars only. As a result, the total elastic strain energy U_{seg} of the multiple bars in Figure 4.3b is $U_{seg} = P_1^2 / (2k_1) + \cdots + P_n^2 / (2k_n)$, while the elastic strain energy U_0 of the single solid bar in Figure 4.3a is $U_0 = P^2 / [2(k_1 + k_2 + \cdots + k_n)]$. Calculating the total elastic strain energy of the system configuration of multiple bars in Figure 4.3b has been done by applying Equation 4.12 n times, where n is the number of segmented bars. The same Equation 4.12 for the elastic strain energy holds for every single bar with a smaller cross section.

The elastic strain energies of the two system configurations can now be compared through inequality (4.1). Inequality (4.1) can be rewritten as

$$\frac{P^2}{2(k_1 + k_2 + \cdots + k_n)} \leq \frac{P_1^2}{2k_1} + \cdots + \frac{P_n^2}{2k_n} \tag{4.13}$$

The left-hand side of inequality (4.13) corresponds to the system configuration in Figure (4.3a), and can be interpreted as accumulated elastic strain energy due to a loading force P acting on a single bar with stiffness $k = k_1 + k_2 + \cdots + k_n$. The right-hand side of inequality (4.13) corresponds to the system configuration in Figure (4.3b), and can be interpreted as accumulated elastic strain energy, resulting from segmenting the original force P into smaller forces and applying the smaller forces P_i to the individual bars with stiffness k_i into which the original bar has been segmented (Todinov, 2020d).

Inequality (4.13) now predicts that *the accumulated elastic strain energy from a loading force acting on a single bar is smaller than the accumulated elastic strain energy, resulting from segmenting the force into smaller forces loading the individual bars into which the original bar has been segmented.*

It needs to be pointed out that not every load segmentation of the original bar leads to an increase of the accumulated elastic strain energy. If all loaded bars have the same displacement, no increase in the accumulated elastic strain energy is present. Indeed, in inequality (4.13), $\delta_i = P_i / k_i$ is the displacement of the ith bar under load P_i. If all bar displacements are equal ($\delta_1 = \delta_2 = ... = \delta_n = \delta$), it is easy to verify that equality is attained in (4.13). Indeed, the right-hand side of inequality (4.13) gives $\frac{1}{2}\left(\frac{P_1^2}{k_1} + \cdots + \frac{P_n^2}{k_n}\right) = (1/2)\delta(P_1 + P_2 + \cdots + P_n) = (1/2)\delta \times P$.

Since $P / (k_1 + \cdots + k_n) = \delta$ and for the left-hand side of inequality (4.13) $P^2 / [2(k_1 + \cdots + k_n)] = (1/2)\delta \times P$ holds, equality is, indeed, attained in inequality (4.13). Note that the loading in Figure 4.3a and b differs significantly. In Figure 4.3b, the bar and the force have been split into smaller bars and forces, and each individual bar has been loaded independently. As a result, the stresses and displacements across the segmented bars are no longer uniform.

Inequality (4.13) holds irrespective of the magnitude of the separate forces into which the initial force P has been segmented and the cross sections of the individual bars. The inequality provides a method of increasing the capacity for absorbing elastic strain energy upon dynamic loading.

This is an unexpected result. To illustrate its validity, consider an example of a steel bar with cross section 10 mm×20 mm and length 1 m, loaded with a force of 50 kN. The Young's modulus of the steel is 210 GPa and the yield strength of the material is 650 MPa.

According to equation 4.12, the accumulated elastic strain energy in the bar with stiffness $k = EA / L$, loaded with a force $P_0 = 50$ kN, is given by

$$U_0 = \frac{P_0^{\,2}}{2k} = \frac{(50\times 10^3)^2}{2\times[(210\times 10^9\times 20\times 10\times 10^{-6})/1]} = 29.76J \tag{4.14}$$

Now, suppose that the bar has been segmented into two bars with cross sections 10 mm×10 mm and length 1 m, and the force P_0 has been segmented into two *unequal* forces with magnitudes 40 and 10 kN, applied to the individual bars. In this case, the accumulated elastic strain energy in the bars is

$$U_1 = \frac{P_1^2}{2k_1} + \frac{P_2^2}{2k_2} = \frac{(40\times 10^3)^2}{2\times[210\times 10^9\times 10\times 10\times 10^{-6}/1]} + \frac{(10\times 10^3)^2}{2\times[210\times 10^9\times 10\times 10\times 10^{-6}/1]}$$

$$= 38.1 + 2.38 = 40.48\,\text{J} \tag{4.15}$$

This is about 36% larger than the elastic strain energy of 29.76 J accumulated in the single bar.

Increasing the capability of accumulating elastic strain energy is important not only to prevent failure during dynamic loading but also in cases where more elastic strain energy needs to be stored.

If the bar had been segmented into two bars with cross sections 8 mm×10 mm and 12 mm×10 mm and the force had been segmented into two forces with magnitudes 20 and 30 kN, the accumulated elastic strain energy characterising the segmented bar and loading would be equal to the elastic strain energy characterising the original bar and loading:

$$U_1 = \frac{P_1^2}{2k_1} + \frac{P_2^2}{2k_2} = \frac{(20\times 10^3)^2}{2\times[210\times 10^9\times 8\times 10\times 10^{-6}/1]} + \frac{(30\times 10^3)^2}{2\times[210\times 10^9\times 12\times 10\times 10^{-6}/1]} = 29.76\,\text{J} \tag{4.16}$$

This is because the displacements of the segmented bars are equal (the ratios $P_1 / k_1 = P_2 / k_2 = \delta$ are the same) and, according to the properties of inequality (4.13), in this case, equality is attained.

If the bar had been segmented into two bars with cross sections 10mm × 10mm and the force had been segmented into two equal forces with magnitude 25 kN, again, the accumulated elastic strain energy in the bar would be the same:

$$U_1 = \frac{P_1^2}{2k_1} + \frac{P_2^2}{2k_2} = 2 \times \frac{(25 \times 10^3)^2}{2 \times [210 \times 10^9 \times 10 \times 10 \times 10^{-6} / 1]} = 29.76\text{J} \quad (4.17)$$

because the displacements of the bars $P_1 / k_1 = P_2 / k_2 = \delta$ are the same.

In order to obtain any advantage from inequality (4.13), *asymmetry must be present in the system* so *that the displacements of the bars are not equal*: $\delta_1 = P_1 / k_1 \neq \delta_2 = P_2 / k_2$. Existence of asymmetry is absolutely essential for increasing the accumulated elastic strain energy through segmentation. Segmented bars experiencing the same displacement do not yield an increase of the amount of stored elastic strain energy.

The asymmetry requirement to achieve increased elastic energy accumulation is rather counterintuitive and makes this result difficult to obtain by alternative means bypassing inequality (4.1).

Consider the sequence $\{P_1^2, P_2^2, \ldots, P_n^2\}$ and the sequence $\{1/(2k_1), 1/(2k_2), \ldots, 1/(2k_n)\}$. The dot product of these sequences is the right-hand side of inequality (4.13). According to the rearrangement inequality, the dot product of two sequences is maximum if they are similarly ordered, for example, if $P_1^2 \leq P_2^2 \leq \ldots \leq P_n^2$ and $1/(2k_1) \leq 1/(2k_2) \leq, \ldots, \leq 1/(2k_n)$. For the first sequence, it can be shown that if $P_1^2 \leq P_2^2 \leq \ldots \leq P_n^2$, then $P_1 \leq P_2 \leq \ldots \leq P_n$. Indeed, from the basic properties of inequalities, for positive P_i and P_j from $P_i^2 \leq P_j^2$ it follows that $P_i \leq P_j$.

For the second sequence, it is easy to see that $1/k_1 \leq 1/k_2 \leq, \ldots, \leq 1/k_n$ only if $k_1 \geq k_2 \geq \ldots \geq k_n$. For any other permutation of the sequences, for example, $\{P_{k1}, P_{k2}, \ldots, P_{kn}\}$ and $\{k_{s1}, k_{s2}, \ldots, k_{sn}\}$, the next inequality is fulfilled:

$$\frac{P_1^2}{2k_1} + \cdots + \frac{P_n^2}{2k_n} \geq \frac{P_{k1}^2}{2k_{s1}} + \cdots + \frac{P_{kn}^2}{2k_{sn}} \quad (4.18)$$

As a result, the right-hand side of inequality (4.13) is maximised if the ordered in ascending order segmented loads $P_1 \leq P_2 \leq \ldots \leq P_n$ are paired with the ordered in descending order stiffness values: $k_1 \geq k_2 \geq \ldots \geq k_n$.

4.5.2 Increasing the Accumulated Elastic Strain Energy for Components Loaded in Bending

It is a well-known result from mechanics of materials that the accumulated elastic strain energy U in a cantilever elastic beam (Figure 4.4a) with rectangular cross section $b \times h$ and length L is given by the equation:

$$U = \frac{1}{2} P \times f \quad (4.19)$$

where P [N] denotes the loading force and f [m] is the deflection of the beam at the point of application of the concentrated force P.

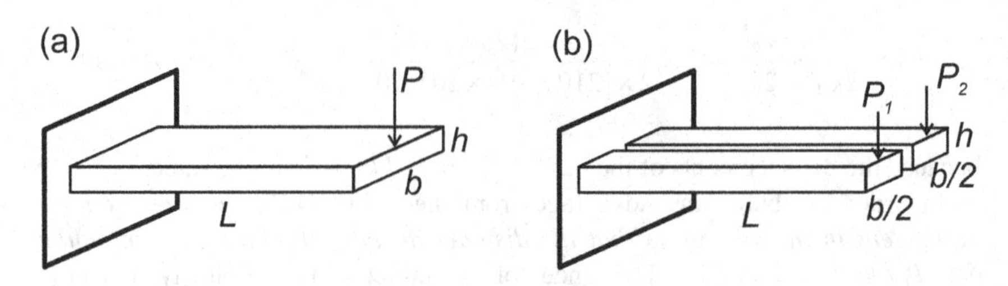

FIGURE 4.4 (a) Cantilever beam loaded with a single concentrated force P; (b) segmented cantilever beam loaded with two forces P_1 and P_2 whose sum is equal to the original force P. (the segmented beams are loaded independently and do not necessarily have the same displacements and stresses).

From mechanics of materials (Gere and Timoshenko, 1999; Hearn, 1985), for a cantilever beam with rectangular cross section, the deflection f can be determined from

$$f = \frac{PL^3}{3EI} \tag{4.20}$$

where E is the Young's modulus of the material and $I = bh^3/12$ is the second moment of area of the beam. Substituting I in equation (4.20) gives

$$f = \frac{4PL^3}{Ebh^3} \tag{4.21}$$

According to equation 4.19, the accumulated elastic strain energy U in the cantilever elastic beam is then given by

$$U = \frac{2P^2L^3}{Ebh^3} \tag{4.22}$$

By introducing the variable $k = \dfrac{Ebh^3}{4L^3}$, standing for the *flexural stiffness* of the cantilever beam, equation 4.19 for the accumulated elastic strain energy can be written as

$$U = \frac{P^2}{2k} \tag{4.23}$$

Suppose that the load P and the cantilever beam have been segmented into n loads $P_1, P_2,\ldots, P_n \left(\sum_{i=1}^{n} P_i = P\right)$ and n beams with the same thickness h and smaller widths $b_1, b_2,\ldots, b_n \left(\sum_{i=1}^{n} b_i = b\right)$. The flexural stiffness k of the original beam is then equal to the sum of the flexural stiffness values characterising the smaller cantilever beams: $k = k_1 + k_2 + \cdots + k_n$, where $k_i = \dfrac{Eb_ih^3}{4L^3}$.

According to inequality (4.1),

$$\frac{P_1^2}{2k_1} + \frac{P_2^2}{2k_2} + \cdots + \frac{P_n^2}{2k_n} \geq \frac{(P_1 + P_2 + \cdots + P_n)^2}{2(k_1 + k_2 + \cdots + k_n)} \tag{4.24}$$

Again, the left- and right-hand side of inequality (4.24) can be interpreted as 'accumulated elastic energy' characterising the alternative design options in Figure 4b and 4a, both experiencing the same overload with magnitude P. The right- and left-hand side of inequality (4.24) can be interpreted as 'accumulated elastic energy' characterising the alternative design options in Figure 4a and 4b, both experiencing the same overload with magnitude P. The right-hand side of inequality (4.24) can be interpreted as the accumulated elastic strain energy due to a load P acting on a cantilever beam with flexural stiffness $k = k_1 + k_2 + \cdots + k_n$ (Figure 4.4a). The left-hand side of inequality (4.24) can be interpreted as accumulated elastic strain energy from loads P_i $\left(\sum_i P_i = P\right)$, resulting from segmenting the original load P into smaller loads and applying the smaller loads to the smaller individual cantilever beams with flexural stiffness k_i into which the original beam has been segmented (Figure 4.4b).

Inequality (4.24) predicts that *the accumulated elastic strain energy due to a load acting on a cantilever beam is smaller than the accumulated strain energy from loads resulting from segmenting the load into smaller loads and applying the smaller loads on the cantilever beam segments.*

Similar to the previous example related to bars, not every segmentation of the beam and the load leads to an increase of the accumulated elastic strain energy. If the smaller cantilever beams have the same displacement under their loads, no increase in the accumulated strain energy will be present. Indeed, in inequality (4.24), the value $f_i = P_i / k_i$ is the displacement of the ith cantilever beam under load P_i. If all beam displacements are equal ($f_1 = f_2 = ... = f_n = f$), equality is attained in inequality (4.24). Indeed, the left-hand side of inequality (4.24) gives $\frac{P_1^2}{2k_1} + \cdots + \frac{P_n^2}{2k_n} = (f/2)(P_1 + P_2 + \cdots + P_n) = (f/2) \times P$. Since, $P/(k_1 + \cdots + k_n) = f$, the right-hand side of inequality (4.24) gives $P^2 / [2(k_1 + ... + k_n)] = (f/2) \times P$, hence equality is indeed attained in inequality (4.24).

Again, the requirement for asymmetry to achieve the effect of increased elastic strain energy is rather counterintuitive and makes this result difficult to obtain by alternative means bypassing the use of inequality (4.1).

The meaningful interpretation of the abstract inequality (4.1) helped find overlooked properties in the mature field of mechanical engineering. No such results have been reported in modern comprehensive textbooks in the area of mechanical engineering and stress analysis (Collins, 2003; Norton, 2006; Pahl et al., 2007; Childs, 2014; Budynas, 1999; Budynas and Nisbett, 2015; Mott et al., 2018; Gullo and Dixon, 2018; Gere and Timoshenko, 1999).

New results can always be obtained by interpretation of the inequalities based on sub-additive functions as long as the variables in the sub- or super-additive functions represent additive quantities. The generated new knowledge can be used for optimising various systems and processes in any area of science and technology.

5 Enhancing Systems Performance by Interpretation of Other Algebraic Inequalities Based on Sub-Additive and Super-Additive Functions

5.1 INCREASING THE ABSORBED KINETIC ENERGY DURING A PERFECTLY INELASTIC COLLISION

Another special case of the general sub-additive function (2.38) is the algebraic inequality

$$\frac{a_1^2}{a_1+b_1}+\frac{a_2^2}{a_2+b_2}+\cdots+\frac{a_n^2}{a_n+b_n}\geq\frac{a^2}{a+b} \tag{5.1}$$

where both controlling factors $a = a_1 + a_2 + \cdots + a_n$ and $b = b_1 + b_2 + \cdots + b_n$ are additive positive quantities. Inequality (5.1) can be proved by induction.

For $n = 2$, inequality (5.1) becomes

$$\frac{a_1^2}{a_1+b_1}+\frac{a_2^2}{a_2+b_2}\geq\frac{(a_1+a_2)^2}{a_1+b_1+a_2+b_2} \tag{5.2}$$

The special case (5.2) can be proved by showing that $\frac{a_1^2}{a_1+b_1}+\frac{a_2^2}{a_2+b_2}-\frac{(a_1+a_2)^2}{a_1+b_1+a_2+b_2}\geq 0$. Since

$$\frac{a_1^2}{a_1+b_1}+\frac{a_2^2}{a_2+b_2}-\frac{(a_1+a_2)^2}{a_1+b_1+a_2+b_2}=\frac{(a_1b_2-a_2b_1)^2}{(a_1+b_1)(a_2+b_2)(a_1+b_1+a_2+b_2)} \tag{5.3}$$

DOI: 10.1201/9781003199830-5

is a non-negative number, the special case (5.2) has been proved. Now, inequality (5.1) can be proved for $n = 3$.

Adding the term $\frac{a_3^2}{a_3 + b_3}$ to both sides of the already proved inequality (5.2) results in

$$\frac{a_1^2}{a_1 + b_1} + \frac{a_2^2}{a_2 + b_2} + \frac{a_3^2}{a_3 + b_3} \geq \frac{(a_1 + a_2)^2}{a_1 + a_2 + b_1 + b_2} + \frac{a_3^2}{a_3 + b_3}$$

Introducing the new positive variables $p = a_1 + a_2$, $q = b_1 + b_2$ and using the already proved case for $n = 2$ result in

$$\frac{p^2}{p+q} + \frac{a_3^2}{a_3 + b_3} \geq \frac{(p + a_3)^2}{p + q + a_3 + b_3} = \frac{(a_1 + a_2 + a_3)^2}{a_1 + b_1 + a_2 + b_2 + a_3 + b_3} \quad (5.4)$$

As a result,

$$\frac{a_1^2}{a_1 + b_1} + \frac{a_2^2}{a_2 + b_2} + \frac{a_3^2}{a_3 + b_3} \geq \frac{(a_1 + a_2 + a_3)^2}{a_1 + a_2 + a_3 + b_1 + b_2 + b_3} \quad (5.5)$$

so the case $n = 3$ has also been proved. Continuing this reasoning proves inequality (5.1) inductively, for any $n > 3$.

The left- and right-hand side of inequality (5.1), multiplied by an appropriate factor can be interpreted as kinetic energy after a perfectly inelastic collision.

Indeed, if an object with mass a, moving horizontally (parallel to the ground) with velocity v_0, collides with a stationary object with mass b and the collision is perfectly inelastic, a single object with mass $a + b$ is formed after the collision, moving with velocity v (Figure 5.1a).

According to the law of conservation of the linear momentum, the sum of the two momenta before collision is equal to their sum after collision

$$av_0 + 0 = (a + b)v$$

from which the velocity v of the two objects after the inelastic collision is given by

$$v = \frac{av_0}{a + b} \quad (5.6)$$

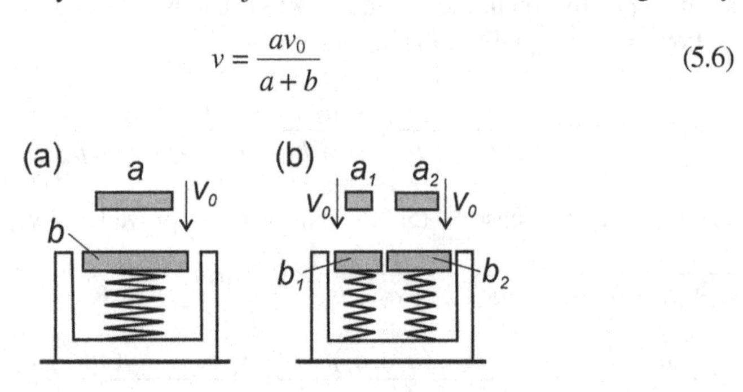

FIGURE 5.1 (a) A perfect inelastic collision between an object with mass a, moving with constant velocity v_0, and a stationary object with mass b. (b) A perfect inelastic collision between a pair of objects with masses a_1 and a_2, moving with constant velocity v_0, and a pair of stationary objects with masses b_1 and b_2.

The kinetic energy of the system after the inelastic impact is, therefore, equal to

$$E_k = \frac{1}{2}(a+b)v^2 = \frac{a^2 v_0^2}{2(a+b)}.$$

Therefore, the right-hand side of inequality (5.1) multiplied by the constant $v_0^2/2$ gives the kinetic energy after the inelastic collision of objects with masses a and b where v_0 is velocity of the moving object.

Suppose that the object with mass a has been segmented into two objects with masses a_1 and a_2 ($a = a_1 + a_2$) and the object with mass b has also been segmented into two objects with masses b_1 and b_2 ($b = b_1 + b_2$) (Figure 5.1b). Multiplying both sides of inequality (5.2) by the factor $v_0^2 / 2$ yields

$$\frac{a_1^2 v_0^2}{2(a_1+b_1)} + \frac{a_2^2 v_0^2}{2(a_2+b_2)} \geq \frac{(a_1+a_2)^2 v_0^2}{2(a_1+b_1+a_2+b_2)} \tag{5.7}$$

which makes the resultant inequality (5.7) interpretable. Both sides of inequality (5.7) effectively describe the same output (total kinetic energy) associated with the design options in Figure 5.1a and 5.1b.

The left-hand side of inequality (5.7) is the total kinetic energy after a perfectly inelastic collision of two objects with masses a_1 and a_2 moving with velocity v_0 towards two stationary objects with masses b_1 and b_2 (Figure 5.1b). The right-hand side of inequality (5.7) is the kinetic energy after a perfectly inelastic collision of a single object with mass equal to the combined mass of the two objects, moving with velocity v_0 towards a single object with mass equal to the combined mass of the stationary objects (Figure 5.1a).

Inequality (5.7) predicts that the aggregation of the objects colliding perfectly inelastically results in a smaller total kinetic energy after the inelastic collision.

Now consider the total kinetic energy $(1/2)a_1 v_0^2 + (1/2)a_2 v_0^2$ of the two moving objects before the inelastic collision. It is equal to the kinetic energy $(1/2)a v_0^2$ before the collision of the single moving object:

$$(1/2)a_1 v_0^2 + (1/2)a_2 v_0^2 = (1/2)a v_0^2 \tag{5.8}$$

Multiplying inequality (5.7) by '–1' and adding to both sides, equation (5.8) results in the inequality

$$\left(\frac{a_1 v_0^2}{2} - \frac{a_1^2 v_0^2}{2(a_1+b_1)}\right) + \left(\frac{a_2 v_0^2}{2} - \frac{a_2^2 v_0^2}{2(a_2+b_2)}\right) \leq \frac{(a_1+a_2)v_0^2}{2} - \frac{(a_1+a_2)^2 v_0^2}{2(a_1+a_2+b_1+b_2)} \tag{5.9}$$

Both sides of inequality (5.9) represent the same chosen output: the absorbed kinetic energy during inelastic collision, characterising the design options in Figure 5.1a and 5.1b. The left-hand side of inequality (5.9) is the absorbed kinetic energy during the inelastic collision of the two pairs of objects, while the right-hand side is the absorbed kinetic energy during the inelastic collision of the single objects.

Inequality (5.7) yielded the interesting prediction that a perfectly inelastic impact between single objects is associated with a greater amount of absorbed kinetic energy compared to the perfectly inelastic collision between the parts of the segmented objects. Aggregating objects, therefore, increases the absorbed kinetic energy during inelastic collision. The prediction obtained from inequality (5.9) can be applied as a basis of a strategy for mitigating inelastic impacts between moving objects and immobile obstacles.

It may seem that the systems in Figure 5.1 are two different systems. In fact, Figures 5.1a and 5.1b are alternative designs of the same shock-absorbing system. Aggregating the impacting masses increases the absorbed kinetic energy during inelastic collision and in this respect, the design option in Figure 5.1a should be preferred.

It needs to be pointed out that aggregation does not always achieve an increase of the absorbed kinetic energy for the system in Figure 5.1. Thus, if $a_1 / b_1 = a_2 / b_2$, the right-hand side of equation (5.3) is zero, equality is attained in inequality (5.2) and no increase in the absorbed kinetic energy is present after the inelastic collision.

Asymmetry must be present for an increase of the absorbed kinetic energy to occur. Again, the requirement for asymmetry to increase the absorbed kinetics energy is rather counter-intuitive and makes this prediction difficult to make by alternative means bypassing inequality (5.1).

5.2 RANKING THE STIFFNESS OF ALTERNATIVE MECHANICAL ASSEMBLIES BY MEANINGFUL INTERPRETATION OF AN ALGEBRAIC INEQUALITY

A special case of the general super-additive function (2.39) is the algebraic inequality

$$\frac{a_1 b_1}{a_1 + b_1} + \frac{a_2 b_2}{a_2 + b_2} + \cdots + \frac{a_n b_n}{a_n + b_n} \le \frac{(a_1 + a_2 + \cdots + a_n)(b_1 + b_2 + \cdots + b_n)}{(a_1 + a_2 + \cdots + a_n) + (b_1 + b_2 + \cdots + b_n)} \tag{5.10}$$

where both controlling factors $a = a_1 + a_2 + \cdots + a_n$ and $b = b_1 + b_2 + \cdots + b_n$ are additive positive quantities.

This inequality can be proved by combining direct manipulation which reduces the inequality to an inequality already considered in Chapter 2. The left-hand side of inequality (5.10) can be presented as

$$\sum_{i=1}^{n} \frac{a_i b_i}{a_i + b_i} = \frac{a_1 b_1}{a_1 + b_1} - a_1 + \frac{a_2 b_2}{a_2 + b_2} - a_2 + \cdots + \frac{a_n b_n}{a_n + b_n} - a_n + (a_1 + \cdots + a_n) \tag{5.11}$$

Conducting the subtractions $\frac{a_i b_i}{a_i + b_i} - a_i$ transforms equality (5.11) into the equality

$$\sum_{i=1}^{n}\frac{a_i b_i}{a_i+b_i} = (a_1+\cdots+a_n) - \frac{a_1^2}{a_1+b_1} - \frac{a_2^2}{a_2+b_2} - \cdots - \frac{a_n^2}{a_n+b_n} \tag{5.12}$$

According to the Bergström inequality,

$$\frac{a_1^2}{a_1+b_1} + \frac{a_2^2}{a_2+b_2} + \cdots + \frac{a_n^2}{a_n+b_n} \geq \frac{(a_1+\cdots+a_n)\times(a_1+\cdots+a_n)}{(a_1+\cdots+a_n)+(b_1+\cdots+b_n)}$$

Consequently, after the substitution in the right-hand side of equality (5.12), the next inequality follows:

$$\sum_{i=1}^{n}\frac{a_i b_i}{a_i+b_i} \leq (a_1+\cdots+a_n) - \frac{(a_1+\cdots+a_n)\times(a_1+\cdots+a_n)}{(a_1+\cdots+a_n)+(b_1+\cdots+b_n)}$$

$$= \frac{(a_1+\cdots+a_n)\times(b_1+\cdots+b_n)}{(a_1+\cdots+a_n)+(b_1+\cdots+b_n)}$$

which completes the proof of inequality (5.10).

Inequality (5.10) has an interesting interpretation. Suppose that there are n pairs of elastic elements A_i, B_i, and a_i stands for the stiffness of elastic element A_i while b_i stands for the stiffness of elastic element B_i, $i = 1,\ldots,n$. Noticing that, in this case, $\frac{a_i b_i}{a_i+b_i}$ is the equivalent stiffness of the elastic elements A_i and B_i connected in series, inequality (5.10) yields an interesting prediction.

The equivalent stiffness of n pairs of elastic elements A_i, B_i working in parallel, where the elastic elements A_i, B_i are connected in series (Figure 5.2a), is smaller than the equivalent stiffness of the system in Figure 5.2b where all A_i and B_i elements are connected in parallel and the two assemblies are connected in series.

However, for $b_i / a_i = k$, $i = 1,2,\ldots,n$, it can be shown that the two assemblies have the same stiffness and are equivalent.

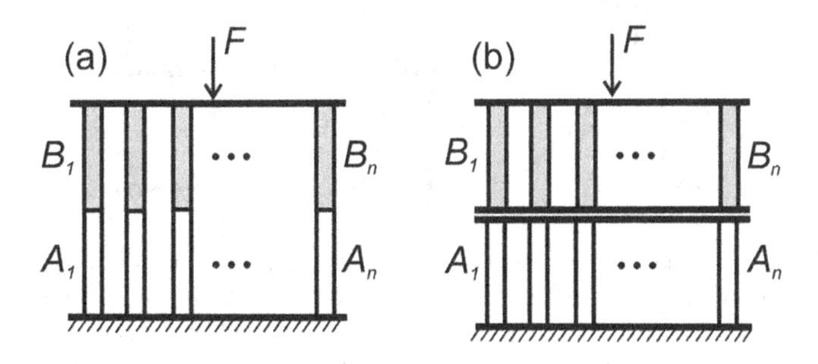

FIGURE 5.2 Two alternative assemblies with different arrangement of the elastic components (a) without bracing plates in the middle, (b) with bracing plates in the middle connecting elastic elements working in parallel.

Indeed, for $b_i = ka_i$, the substitution in the left-hand side of inequality (5.10) yields:

$$\frac{a_1 k}{1+k} + \frac{a_2 k}{1+k} + \cdots + \frac{a_n k}{1+k} = \frac{(a_1 + a_2 + \cdots + a_n)k}{1+k} \tag{5.13}$$

The substitution in the right-hand side of inequality (5.10) results in:

$$\frac{(a_1 + a_2 + \cdots + a_n)(b_1 + b_2 + \cdots + b_n)}{(a_1 + a_2 + \cdots + a_n) + (b_1 + b_2 + \cdots + b_n)} = \frac{(a_1 + a_2 + \cdots + a_n)^2 k}{(a_1 + a_2 + \cdots + a_n) + k(a_1 + a_2 + \cdots + a_n)}$$

$$= \frac{(a_1 + a_2 + \cdots + a_n)k}{1+k} \tag{5.14}$$

The right-hand sides of equalities (5.13) and (5.14) are identical; therefore, the left-hand sides are also identical. The two assemblies are equivalent (have the same stiffness).

The important conclusion from the interpretation of inequality (5.10) is that the increase of stiffness associated with the assembly in Figure 5.2b *is lost completely if the pairs are characterised by the same ratio of the stiffness values of the individual elastic elements.*

This is a counter-intuitive statement which can be demonstrated on two pairs of elastic elements A_1, B_1, A_2, B_2 with the same length in unloaded state and stiffness: $a_1 = 1{,}400$ N/m, $b_1 = 800$ N/m, $a_2 = 1{,}050$ N/m $b_2 = 600$ N/m, respectively (Figure 5.3a and b).

Because $a_1 / b_1 = a_2 / b_2 = 1.75$,

$$\frac{a_1 b_1}{a_1 + b_1} + \frac{a_2 b_2}{a_2 + b_2} = \frac{(a_1 + a_2)(b_1 + b_2)}{(a_1 + a_2) + (b_1 + b_2)} = 890.91 \text{N/m}$$

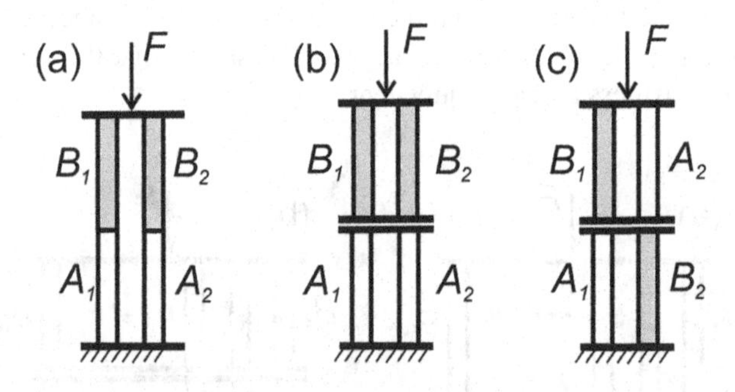

FIGURE 5.3 Alternative assemblies with different arrangement of the elastic components: (a) two pairs of elastic components connected in parallel in which two elastic components of types A and B are connected in series; (b) two elastic assemblies connected in series; one of the assemblies includes two elastic components of type A connected in parallel and the second assembly includes two elastic components of type B connected in parallel; (c) two elastic assemblies connected in series. Each of the assemblies includes two elastic components of type A and type B connected in parallel.

and the assemblies in Figure 5.3a and b have the same stiffness. However, for the pairs of elastic elements $a_1 = 1{,}400\,\text{N/m}$, $b_1 = 800\,\text{N/m}$ and $a_2 = 600\,\text{N/m}$, $b_2 = 1{,}050\,\text{N/m}$ (Figure 5.3c), the stiffness values in the pairs are no longer proportional: $a_1 / b_1 = 1.75 \neq a_2 / b_2 = 0.57$.

As a result, the left- and right-hand side of inequality (5.10) are not equal:

$$\frac{a_1 b_1}{a_1 + b_1} + \frac{a_2 b_2}{a_2 + b_2} = 890.91\,\text{N/m} < \frac{(a_1 + a_2)(b_1 + b_2)}{(a_1 + a_2) + (b_1 + b_2)} = 961\,\text{N/m}$$

and the result is an increase of the stiffness for the assembly in Figure 5.3c compared to the assemblies in Figure 5.3a and b. Inequality (5.10) can be applied as basis for a strategy for increasing the stiffness of mechanical assemblies similar to the ones in Figure 5.2a and 5.2b.

5.3 INTERPRETATION OF INEQUALITIES BASED ON SINGLE-VARIABLE SUPER- AND SUB-ADDITIVE FUNCTIONS

5.3.1 General Inequalities Based on Single-Variable Super- and Sub-Additive Functions

Consider an additive quantity x that has been segmented into a number of non-negative segments $x_1, x_2, \ldots, x_n$ $(x = x_1 + x_2 + \cdots + x_n)$. The output associated with each of the segments x_i is given by $f(x_i)$. According to Section 2.2.3, if the function $f(x)$ is a *sub-additive* function of the additive quantity x, for the parts $x_1, x_2, \ldots, x_n$ into which the additive quantity x has been segmented, the next inequality holds

$$f(x_1 + x_2 + \cdots + x_n) \leq f(x_1) + f(x_2) + \cdots + f(x_n) \tag{5.15}$$

for any set of segments $x_1, x_2, \ldots, x_n$.

According to Section 2.2.3, if the function $f(x)$ is a *super-additive* function of the additive quantity x, for the parts $x_1, x_2, \ldots, x_n$ into which the additive quantity x has been segmented, the next inequality holds

$$f(x_1 + x_2 + \cdots + x_n) \geq f(x_1) + f(x_2) + \cdots + f(x_n) \tag{5.16}$$

for any set of $x_1, x_2, \ldots, x_n$.

Inequalities (5.15) and (5.16) have important potential applications in optimising processes. Suppose that the function $f(x)$ measures the effect/output of a particular additive quantity x, the variables x_i (i=1,...,n) denote the different segments of the quantity x, and the effects $f(x_i)$ are also additive quantities. The inequalities (5.15) and (5.16) then provide the unique opportunity to increase the effect of the additive quantity x by segmenting or aggregating it, depending on whether the function $f(x)$ is concave or convex. If the function $f(x)$ is concave, with a domain $[0, \infty)$ and range $[0, \infty)$, the function is sub-additive and segmenting the extensive quantity results in a larger output. If the function $f(x)$ is convex, with a domain

$[0,\infty)$, and if $f(0) \le 0$, the function is super-additive and aggregating the additive quantity results in a larger output (see Chapter 2, Section 2.2.3).

Inequalities (5.15) and (5.16) have a universal application in science and technology as long as x_i and the terms $f(x_i)$ are additive quantities and have meaningful interpretation.

The use of inequalities based on single-variable sub- and super-additive functions in process optimisation will be illustrated by power-law dependences involving a single controlling factor. Power functions are widespread (Andriani and McKelvey, 2007; Newman, 2007; Easley and Kleinberg, 2010). Many additive quantities can be approximated very well by power functions of the type

$$y = ax^p \tag{5.17}$$

where a and p are constants ($a \ne 0; p > 0$), x is the controlling factor ($x \ge 0$) and y is the output quantity.

The power functions given by equation 5.17 are encountered frequently in mathematical modelling, where y stands for frequency, energy, power, force, damage, profit, pollution, etc. Power laws appear, for example, in cases where positive feedback loops are determining the output. At the heart of these positive feedback loops is often the *preferential attachment*, according to which a commodity is distributed according to how much commodity is already present.

Depending on whether the power p in equation (5.17) is smaller or greater than unity, sub- or super-additive inequalities can be used to perform segmentation or aggregation of the additive controlling factor in order to attain enhanced performance.

The additive output quantity y can be a convex function of the controlling factor x ($x \ge 0$) depending on whether the second derivative $d^2y/dx^2 = ap(p-1)x^{p-2}$ with respect to x is positive or negative. This depends on whether the power p is greater or smaller than 1. If $p > 1$, $d^2y/dx^2 \ge 0$, and the power function y is convex. If $0 < p < 1$, $d^2y/dx^2 \le 0$, and the power function y is concave. If $p < 0$, $d^2y/dx^2 = ap(p-1)x^{p-2} > 0$, and the power function y is convex.

Suppose that the output function y is of the type presented by equation (5.17). For a controlling factor x varying in the interval $[0,\infty)$ the function (5.17) is convex if $p > 1$ and since $y(0) = 0$, the following super-additive inequality holds:

$$a(x_1 + x_2 + \cdots + x_n)^p \ge ax_1^p + ax_2^p + \cdots + ax_n^p \tag{5.18}$$

This inequality means that aggregating the additive factor x yields a larger total effect.

If $0 < p < 1$, the power function (5.17) is concave, and since $y(0) = 0$ the following sub-additive inequality holds

$$a(x_1 + x_2 + \cdots + x_n)^p \le ax_1^p + ax_2^p + \cdots + ax_n^p \tag{5.19}$$

which means that segmenting the additive factor x yields a larger total effect.

5.3.2 An Application of Inequality Based on a Super-Additive Function to Minimise the Formation of Brittle Phase during Solidification

Consider an application example involving the formation of undesirable brittle phase, during the solidification of a specimen with spherical shape and volume V. The specimen solidifies from a particular initial temperature of the molten alloy. The quantity z of undesirable brittle phase formed during solidification (which is an additive quantity) is given by the power function

$$z = aV^p \tag{5.20}$$

where a and p ($p > 1$) are constants which depend on the temperature, the nature of the alloy and the shape of the specimen. The controlling factor in the power-law dependence (5.20) is the volume V of the specimen which is an additive quantity.

The formation of brittle phase during solidification compromises the mechanical properties of the specimen and an increased quantity of brittle phase means significantly reduced strength.

The additive quantity z of brittle phase is a convex function of the volume of the specimen because the second derivative with respect to the volume V is positive: $d^2z / dV^2 = ap(p-1)V^{p-2}$. Since function (5.20) is convex, the quantity z varies in the interval $[0,\infty)$ and $z(0)=0$, the function (5.20) is super-additive. Consequently, the inequality

$$a(V_1 + V_2 + \cdots + V_n)^p \geq aV_1^p + aV_2^p + \cdots + aV_n^p \tag{5.21}$$

holds. The right-hand side of inequality (5.21) can be interpreted as the sum of the quantities of brittle phase in n specimens with volumes $V_1, V_2, \ldots, V_n$. The left-hand side of inequality (5.21) can be interpreted as quantity of brittle phase in a single specimen with a similar shape and volume equal to the sum of the volumes $V_1, V_2, \ldots, V_n$ of the smaller specimens.

As a result, inequality (5.21) predicts that segmenting a specimen with volume V into n specimens with a similar shape and volumes $V_1, V_2, \ldots, V_n$, ($V = V_1 + V_2 + \cdots + V_n$) will decrease the amount of unwanted brittle phase during solidification, irrespective of the individual volumes $V_1, V_2, \ldots, V_n$.

5.3.3 An Application of Inequality Based on a Super-Additive Function to Minimise the Drag Force Experienced by an Object Moving through Fluid

This example illustrates using inequalities based on super-additive power functions to reduce the drag force F_d acting on a body with volume V moving through fluid with a constant velocity v. The drag force is an additive property specified by

$$F_d = (1/2)C_d \rho v^2 A \tag{5.22}$$

where C_d is the drag coefficient – a dimensionless number that depends on the shape of the body, ρ [kg/m³] is the density of the fluid, v [m/s] is the velocity of the body and A [m²] is the cross-sectional area of the body. Considering that for a body with volume V, the cross-sectional area is proportional to $V^{2/3}$, the drag force can be presented as a function of the additive property 'volume V' of the body:

$$F_d = aV^{2/3} \tag{5.23}$$

where a is a constant.

The drag force given by equation (5.23) is a concave function because the second derivative with respect to the volume V is negative: $d^2F_d / dV^2 = -2a/(9V^{4/3})$. Because for volume varying in the interval $[0,\infty)$ the drag force varies in the range $[0,\infty)$, the drag force function (5.23) is sub-additive. Consequently, the inequality

$$a(V_1 + V_2 + \cdots + V_n)^{2/3} \le aV_1^{2/3} + aV_2^{2/3} + \cdots + aV_n^{2/3} \tag{5.24}$$

holds. The right-hand side of inequality (5.24) is an additive quantity and can be interpreted as the total drag force acting on n bodies with volumes $V_1, V_2, \ldots, V_n$ moving with the same speed. The left-hand side of inequality (5.24) is also an additive quantity and can be interpreted as the drag force acting on a body with a similar shape and volume equal to the sum of the volumes $V_1, V_2, \ldots, V_n$ of the separate bodies and moving with the same speed.

As a result, inequality (5.24) predicts that aggregating n bodies with volumes $V_1, V_2, \ldots, V_n$ into a single large body with a similar shape and larger volume $V = V_1 + V_2 + \cdots + V_n$, equal to the sum of the volumes of the original bodies, decreases the total drag force, irrespective of the actual volumes $V_1, V_2, \ldots, V_n$.

5.3.4 Light-Weight Designs by Interpretation of an Algebraic Inequality Based on a Single-Variable Sub-Additive Function

Consider the sub-additive algebraic inequality

$$ka^{2/3} \le ka_1^{2/3} + ka_2^{2/3} + \ldots + ka_n^{2/3} \tag{5.25}$$

where k is a positive constant and the controlling factor $a = a_1 + a_2 + \ldots + a_n$ is an additive positive quantity. Because the power function $y = ka^p, 0 < p < 1$, is a concave function, and the range of $y = ka^p$ is $0 \le y \le \infty$ when $0 \le a \le \infty$, the power function $y = ka^p$ is a sub-additive function.

If the variable a in inequality (5.25) is interpreted as magnitude of a load on a cylindrical cantilever beam (Figure 5.4a), then $ka^{2/3}$ can be interpreted

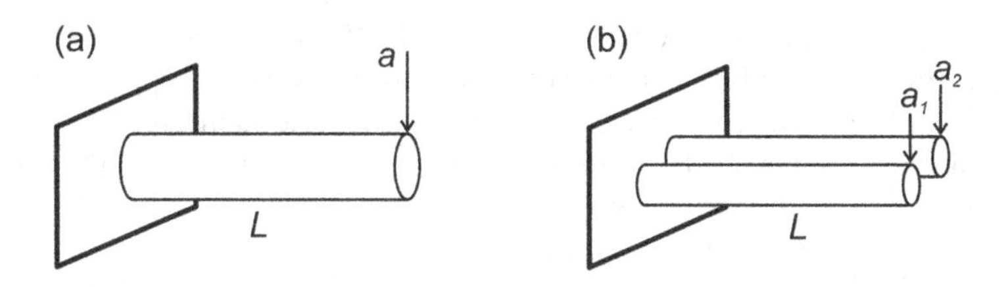

FIGURE 5.4 (a) A cylindrical cantilever beam with length L loaded with a single concentrated force with magnitude a; (b) two identical cantilever beams with smaller radii, loaded with two forces with magnitudes a_1 and a_2 whose sum is equal to the magnitude a of the single force.

as 'minimum volume of the cantilever beam necessary to support a force with magnitude a'. This interpretation corresponds to the design alternative in Figure 5.4a

Indeed, the volume of a cylindrical beam with radius r [m] of the cross section and length L [m] is given by

$$V = \pi r^2 L \tag{5.26}$$

If σ_{cr} is the allowable bending stress, from mechanics of materials (Gere and Timoshenko, 1999):

$$\sigma_{cr} = \frac{M}{I} r \tag{5.27}$$

where $M = aL$ is the loading moment and $I = \pi r^4 / 4$ is the second moment of area for a circular cross section. Substituting the second moment of area I in (5.27) gives

$$\sigma_{cr} = \frac{4aL}{\pi r^3} \tag{5.28}$$

from which, for the minimum radius of the beam capable of supporting the load with magnitude a, $r = \left(\frac{4aL}{\pi \sigma_{cr}} \right)^{1/3}$ is obtained. After substituting r in (5.26), the minimum volume V of the beam necessary for supporting the loading force a becomes

$$V = ka^{2/3} \tag{5.29}$$

where $k = \frac{4^{2/3} \pi^{1/3} L^{5/3}}{\sigma_{cr}^{2/3}}$.

Supporting the load with magnitude a, can also be done by an alternative design option consisting of two cantilever beams with length L, each loaded with forces with magnitudes $a_1 = a_2 = a / 2$ (Figure 5.4b). The two design options with the same function of supporting a total load with magnitude a, depicted in Figure 5.4a and 5.4b, can now be compared through inequality (5.25). The comparison shows that the minimum combined volume of the beams necessary to support the loads a_1 and a_2 can be reduced if, instead of two beams, a single cantilever beam with a larger radius is used, loaded with force $a = a_1 + a_2$ equal to the sum of the loading forces $a_1 = a_2 = a/2$ (Figure 5.4a).

Indeed, according to inequality (5.25):

$$ka^{2/3} < k(a/2)^{2/3} + k(a/2)^{2/3} = ka^{2/3} \times (2/2^{2/3})$$

Since $(2/2^{2/3}) = 1.26$, aggregating the loads and selecting the design option featuring a single cantilever beam with a larger radius, leads to a significant decrease of the minimum volume of material necessary to support the loads.

Lightweight design developed by using algebraic inequalities will be illustrated by a numerical example. Suppose that the cantilever beam has a length of $L = 2.5m$, loaded at the end with a force $a = 12kN$ and the material is wood with allowable bending stress $\sigma_{cr} = 15MPa$.

From $r = \left(\frac{4aL}{\pi\sigma_{cr}}\right)^{1/3} = \left(\frac{4\times12\times10^3\times2.5}{\pi\times15\times10^6}\right)^{1/3} = 0.137$, the volume of wood V necessary to support a load of magnitude $a = 12kN$ is $V = \pi r^2 L = \pi\times0.137^2\times 2.5 = 0.147\ m^3$

From $r_1 = r_2 = \left(\frac{4(a/2)L}{\pi\sigma_{cr}}\right)^{1/3} = \left(\frac{4\times6\times10^3\times2.5}{\pi\times15\times10^6}\right)^{1/3} = 0.108$, the combined volume V' of material of two identical beams needed to support loads of magnitude $a/2 = 6kN$ (the combined load is $a = 12kN$), is $V' = \pi r_1^2 L + \pi r_2^2 L = 2\pi\times0.108^2\times2.5 = 0.183\ m^3$.

The effect from aggregating the loads is greater for a larger number of aggregated beams. Thus, for a supporting structure consisting of three cantilever beams with length *L*, each loaded with force of magnitude $a_1 = a_2 = a_3 = a/3$, according to inequality (5.25):

$$ka^{2/3} < k(a/3)^{2/3} + k(a/3)^{2/3} + k(a/3)^{2/3} = ka^{2/3} \times (3/3^{2/3})$$

Since $(3/3^{2/3}) = 1.44$, aggregating the loads into a single load and the three beams into a single beam with a larger radius, leads to an even greater decrease of the mass of the supporting structure.

To summarise, the same required function (supporting a total load with magnitude *a*) is delivered by two different design options. Inequality (5.25) predicts that the single-beam design option is associated with a smaller volume of material necessary to support the load. For this reason, inequality (5.25) can be used with success for producing light-weight designs in mechanical and structural engineering.

5.3.5 An Application of Inequality Based on a Sub-Additive Function to Maximise the Profit from an Investment

Consider now an application example from economics based on a sub-additive function involving additive quantities. Suppose that the annual profit *z* from an investment in a particular enterprise is given by the power function

$$z = cx^q \tag{5.30}$$

where x is the size of the investment and c and q ($q < 1$) are constants that depend on the particular enterprise. The additive factor in the power-law dependence (5.30) is the size of the investment x. The profit z is also an additive quantity.

For $0 < q < 1$, the profit z is a concave function of the size of investment because the second derivative with respect to the investment x is negative: $d^2z / dx^2 = cq(q-1)x^{q-2} < 0$. Because for investment varying in the interval $[0,\infty)$, the profit varies in the range $[0,\infty)$, function (5.30) giving the profit from the investment is sub-additive. Consequently, inequality (5.19) becomes

$$c(x_1 + x_2 + \cdots + x_n)^q \le cx_1^q + cx_2^q + \cdots + cx_n^q \tag{5.31}$$

Inequality (5.31) predicts that splitting the initial investment x and investing in n parallel enterprises of the same type, will result in larger profit than investing the entire sum x in a single enterprise. The difference in profit can be significant as is shown in the next numerical example. Thus, for a profit dependence

$$f(x) = 15.3x^{0.42}, \tag{5.32}$$

the profit from investing $x = \$10,000$ is $15.3 \times 10,000^{0.42} = \$732.3$. Splitting the investment in two and investing \$5,000 in two parallel enterprises yield profit of magnitude $15.3 \times 5,000^{0.42} + 15.3 \times 5,000^{0.42} = \$1,094.7$ which is by 49.5% larger than the profit obtained from the initial investment.

5.4 INCREASING THE MASS OF SUBSTANCE DEPOSITED DURING ELECTROLYSIS AND AVOIDING OVERESTIMATION OF DENSITY THROUGH INTERPRETATION OF AN ALGEBRAIC INEQUALITY

There are algebraic inequalities, which, although not based on sub- or super-additive functions, still provide the basis for a segmentation of additive factors and increasing their impact. Consider the algebraic inequality

$$\frac{a_1}{x_1} + \frac{a_2}{x_2} + \cdots + \frac{a_n}{x_n} \ge n\frac{a_1 + a_2 + \cdots + a_n}{x_1 + x_2 + \cdots + x_n} \tag{5.33}$$

where $a_1 \ge a_2 \ge \ldots \ge a_n$ and $x_1 \le x_2 \le \ldots \le x_n$ are positive numbers.

Inequality (5.33) is a special case of the general inequality (2.48). The role of the function $f(a,b)$ in inequality (2.48) is played by the function $f(a,x)=a/x$ in inequality (5.33).

This inequality has been proved rigorously in Chapter 2 by sequentially applying the Chebyshev's inequality and the AM–GM inequality.

Presence of asymmetry is vital for inequality (5.33) to hold. For proportional ratios $a_i / x_i = a_j / x_j$, $i = 1,\ldots,n$; $j = 1,\ldots,n$, equality is attained in inequality (5.33).

Indeed, in this case, the left-hand side of inequality (5.33) becomes $\frac{a_1}{x_1}+\frac{a_2}{x_2}+\cdots+\frac{a_n}{x_n}=nr$, and since $a_i = rx_i$, the right-hand side of inequality (5.33) also gives the same value: $n\frac{a_1+a_2+\cdots+a_n}{x_1+x_2+\cdots+x_n}=n\frac{r(x_1+x_2+\cdots+x_n)}{x_1+x_2+\cdots+x_n}=nr$.

Despite that inequality (5.33) is not based on a sub-additive function, it still offers the possibility for a segmentation of additive factors and has a number of interesting potential applications. In order to apply inequality (5.33), the ratios a_i / x_i of the additive quantities a_i and x_i must also be additive quantities.

If a_i / x_i in inequality (5.33) is an additive quantity and a_i, x_i are, in turn, additive quantities, inequality (5.33) provides a mechanism to increase at least n times the effect of the aggregated additive quantities $a=\sum_{i=1}^{n} a_i$ and $x=\sum_{i=1}^{n} x_i$ by segmenting them into smaller parts a_i and x_i, $i = 1,\ldots,n$ and accumulating their individual effects a_i / x_i. Inequality (5.33) has a universal application in science and technology as long as a_i, x_i and the terms a_i / x_i are additive quantities and have meaningful interpretation.

An alternative way of formulating this requirement is to be able to present the additive quantity a_i as a product of the additive quantity p_i and the additive quantity x_i:

$$a_i = p_i \times x_i \tag{5.34}$$

It is important to note that inequality (5.33) can be applied to each of the individual terms a_i / x_i on the left-hand side which, in turn, can be segmented. The result from this recursive segmentation is a further multiplication of the effect from the segmentation.

Inequality (5.33) will be applied to increase the mass of substance deposited on electrodes during electrolysis. The Faraday's first law of electrolysis (Wolfson, 2016) states that the mass m of substance deposited at an electrode in grams is directly proportional to the charge Q in Coulombs:

$$m = Z \times Q \tag{5.35}$$

where Z is a constant of proportionality called electro-chemical equivalent of the substance. It is the mass deposited for a charge of 1 Coulomb.

By using the relationship between charge Q, current I [A] and time t in seconds, equation (5.35) can be rewritten as

$$m = Z \times I \times t \tag{5.36}$$

which gives the mass m of substance deposited at an electrode in grams, by a current of magnitude I, for a time of t seconds.

Consider now a process of electrolysis induced by voltage of magnitude V applied to a cell with resistance R. Since the current I is determined from the Ohm's law

$$I = V / R, \tag{5.37}$$

the mass m_0 of deposited substance is given by

$$m_0 = Z \times (V / R) \times t \tag{5.38}$$

Consider an alternative design, for which the electrolysis process is conducted after segmenting the initial cell with resistance R into two smaller cells with smaller resistances r_1 and r_2 ($r_1 + r_2 = R$; $r_1 \le r_2$). The voltages applied to the cells are with magnitudes V_1 and V_2 ($V_1 \ge V_2$, and $V_1 + V_2 = V$), correspondingly. For the same size of the electrodes and the same ionic solution, the resistance of the solution is proportional to the distance between the electrodes. As a result, the smaller cells, for which the distance between the electrodes has been reduced, are characterised by a smaller electrical resistance. According to inequality (5.33), for $n = 2$, the next inequality holds:

$$\frac{V_1}{r_1} + \frac{V_2}{r_2} \ge 2\frac{V_1 + V_2}{r_1 + r_2} = \frac{2V}{R} \tag{5.39}$$

Multiplying both sides of inequality (5.39) with the positive value $Z \times t$ gives

$$Z(V_1 / r_1)t + Z(V_2 / r_2)t \ge 2Z(V / R)t \tag{5.40}$$

The left-hand side of inequality (5.40) can be interpreted as the sum of masses $m_1 + m_2$ in grams, of substance deposited at the electrodes of the smaller electrolytic cells, by currents of magnitudes $I_1 = V_1 / r_1$ and $I_2 = V_2 / r_2$, for a duration of t seconds. The right-hand side of inequality (5.40) can be interpreted as the mass m_0 of substance deposited at the electrode of the original cell, by current of magnitude $I = V / R$, for the same duration of t seconds.

Inequality (5.40) predicts that, as a result of the cell's segmentation, the mass of deposited substance can be increased more than twice.

Again, the presence of asymmetry in inequality (5.40) is a condition for improved performance. For $V_1 / r_1 = V_2 / r_2$, no increase of the deposited substance is present.

The segmentation of additive quantities through the algebraic inequality (5.33) can be used to achieve systems and processes with superior performance and the algebraic inequality is applicable to any area of science and technology, as long as the additive quantities comply with the simple condition (5.34). Inequality (5.33) can also be rewritten as

$$\frac{1}{n}\left(\frac{a_1}{x_1} + \frac{a_2}{x_2} + \cdots + \frac{a_n}{x_n}\right) \ge \frac{a_1 + a_2 + \cdots + a_n}{x_1 + x_2 + \cdots + x_n} \tag{5.41}$$

Even if the ratios a_i / x_i are non-additive quantities, the left-hand side of the inequality can still be interpreted as an estimate of the intensive quantity $\frac{a_1 + a_2 + \cdots + a_n}{x_1 + x_2 + \cdots + x_n}$ through the average of the segmented intensive quantities a_i / x_i.

The left-hand side of inequality (5.41) can then be interpreted as the average value of the segmented intensive quantities a_i / x_i. Here is an example.

Suppose that a mixture of n distinct incompressible substances has been specified through the masses m_i of the substances and the volumes v_i these substances occupy ($m_1 \geq m_2 \geq ... \geq m_n$; $v_1 \leq v_2 \leq ... \leq v_n$). The density of the mixture is clearly given by $\rho_{mix} = \frac{m_1 + m_2 + \cdots + m_n}{v_1 + v_2 + \cdots + v_n}$. However, if the density of the mixture is estimated by using $\hat{\rho}_{\text{mix}} = \frac{1}{n}\left(\frac{m_1}{V_1} + \frac{m_2}{V_2} + \cdots + \frac{m_n}{V_n}\right)$, this could result in an over-estimation of the real density.

The overestimation can be significant and will be illustrated by the next numerical example. For three distinct substances with masses $m_1 = 125\,\text{g}$, $m_2 = 50\,\text{g}$ and $m_3 = 25\,\text{g}$ and volumes $v_1 = 10\,\text{cm}^3$, $v_2 = 15\,\text{cm}^3$ and $v_3 = 20\,\text{cm}^3$, correspondingly, inequality (5.41) becomes

$$\frac{1}{3}\left(\frac{m_1}{v_1} + \frac{m_2}{v_2} + \frac{m_3}{v_3}\right) \geq \frac{M}{V} \tag{5.42}$$

The average density estimated by the left-hand side of inequality (5.42) is: $(1/3)(m_1/v_1 + m_2/v_2 + m_3/v_3) = (1/3)(125/10 + 50/15 + 25/20) = 5.69\ \text{g/cm}^3$.

This estimated value is significantly larger than the correct value $M / V = 200 / 45 = 4.44\,\text{g/cm}^3$ for the density of the mixture.

5.5 GENERATING NEW KNOWLEDGE ABOUT THE DEFLECTIONS OF ELASTIC ELEMENTS ARRANGED IN SERIES AND PARALLEL

If the variables a_i in inequality (5.33) are interpreted as loads on n elastic elements and x_i are interpreted as the stiffness values of these elastic elements, the terms a_i / x_i can be interpreted as deflections of the elastic elements under these loads. The loads a_i are additive quantities and the stiffness values for elastic elements connected in parallel are also additive quantities.

Suppose also that $a_1 = a_2 = \ldots = a_n = F / n$. If x_i are ordered in ascending order ($x_1 \leq x_2 \leq \ldots \leq x_n$) because $a_1 = a_2 = \ldots = a_n = F/n$, the conditions $a_1 \geq a_2 \geq \ldots \geq a_n$, $x_1 \leq x_2 \leq \ldots \leq x_n$ for the validity of inequality (5.33) are fulfilled.

Inequality (5.33) can then be re-written as

$$\frac{F}{x_1 + x_2 + \cdots + x_n} \leq \frac{1}{n}\left(\frac{F/n}{x_1} + \frac{F/n}{x_2} + \cdots + \frac{F/n}{x_n}\right) = \frac{1}{n^2}\left(\frac{F}{x_1} + \frac{F}{x_2} + \cdots + \frac{F}{x_n}\right) \tag{5.43}$$

The left-hand side of inequality (5.43) can be interpreted as deflection of an assembly of n elastic elements connected in parallel (see Figure 3.1a) and loaded with a force of magnitude F. The right-hand side of inequality (5.43) is the total deflection of the same n elastic elements connected in series and loaded with a single force of the same magnitude F (see Figure 3.1b).

The interpretation of inequality (5.43) states that the deflection of elastic elements connected in parallel is more than n^2 times smaller than the deflection of the same elastic elements connected in series and loaded with a force of the same magnitude.

6 Optimal Selection and Expected Time of Unsatisfied Demand by Meaningful Interpretation of Algebraic Inequalities

6.1 MAXIMISING THE PROBABILITY OF SUCCESSFUL SELECTION FROM SUPPLIERS WITH UNKNOWN PROPORTIONS OF HIGH-RELIABILITY COMPONENTS

The approach based on meaningful interpretation of correct algebraic inequalities can also be applied for ranking alternative decisions. Consider the non-trivial abstract inequalities:

$$x_1^2 + x_2^2 + x_3^2 \geq x_1x_2 + x_2x_3 + x_3x_1 \tag{6.1}$$

$$x_1^3 + x_2^3 + x_3^3 \geq 3x_1x_2x_3 \tag{6.2}$$

$$2(x_1^3 + x_2^3 + x_3^3) \geq x_1^2x_2 + x_1^2x_3 + x_2^2x_1 + x_2^2x_3 + x_3^2x_1 + x_3^2x_2 \tag{6.3}$$

$$2(x_1^4 + x_2^4 + x_3^4) \geq x_1^3x_2 + x_1^3x_3 + x_2^3x_1 + x_2^3x_3 + x_3^3x_1 + x_3^3x_2 \tag{6.4}$$

Inequalities 6.1–6.4 can be proved by invoking the Muirhead's inequality introduced in Chapter 2.

For any set of non-negative numbers $x_1, x_2, \ldots, x_n$, a symmetric sum is defined as $\sum_{\text{sym}} x_1^{a_1} x_2^{a_2} \ldots x_n^{a_n}$ which, when expanded, includes $n!$ terms. Each term is formed by a distinct permutation of the elements of the sequence $a_1, a_2, \ldots, a_n$.

The Muirhead's inequality states that if the sequence $\{a\}$ is majorising sequence $\{b\}$ and $x_1, x_2, \ldots, x_n$ are non-negative, the next inequality holds:

$$\sum_{\text{sym}} x_1^{a_1} x_2^{a_2} \ldots x_n^{a_n} \geq \sum_{\text{sym}} x_1^{b_1} x_2^{b_2} \ldots x_n^{b_n} \tag{6.5}$$

DOI: 10.1201/9781003199830-6

Consider the two non-increasing sequences $a_1 \geq a_2 \geq, \ldots, \geq a_n$ and $b_1 \geq b_2 \geq, \ldots, \geq b_n$ of non-negative real numbers. The sequence $\{a\}$ is said to majorise sequence $\{b\}$ if the following conditions are fulfilled:

$$a_1 \geq b_1;\ a_1 + a_2 \geq b_1 + b_2;\ \ldots;\ a_1 + a_2 + \ldots + a_{n-1} \geq b_1 + b_2 + \ldots + b_{n-1}$$
$$a_1 + a_2 + \cdots + a_{n-1} + a_n = b_1 + b_2 + \cdots + b_{n-1} + b_n \tag{6.6}$$

Consider now the sequences $\{a\} = [2,0,0]$ and $\{b\} = [1,1,0]$. Because the sequence $\{a\} = [2,0,0]$ majorises sequence $\{b\} = [1,1,0]$, inequality (6.7) is obtained.

$$2! \times \left(x_1^2 + x_2^2 + x_3^2\right) \geq 2x_1x_2 + 2x_2x_3 + 2x_3x_1 \tag{6.7}$$

Dividing both sides of inequality (6.7) by 2! transforms inequality (6.7) into inequality (6.1).

Consider now the sequences $\{a\} = [3,0,0]$ and $\{b\} = [1,1,1]$. Because the sequence $\{a\} = [3,0,0]$ majorises sequence $\{b\} = [1,1,1]$, inequality (6.8) is obtained.

$$2! \times \left(x_1^3 + x_2^3 + x_3^3\right) \geq 3! \times x_1x_2x_3 \tag{6.8}$$

Dividing both sides of inequality (6.8) by 2! transforms inequality (6.8) into inequality (6.2).

Next, since the sequence $\{a\} = [3,0,0]$ majorises the sequence $\{b\} = [2,1,0]$, the following inequality also follows immediately from the Muirhead's inequality (6.5):

$$2! \times \left(x_1^3 + x_2^3 + x_3^3\right) \geq x_1^2x_2 + x_1^2x_3 + x_2^2x_1 + x_2^2x_3 + x_3^2x_1 + x_3^2x_2 \tag{6.9}$$

which gives inequality (6.3).

Finally, since the sequence $\{a\} = [4,0,0]$ majorises the sequence $\{b\} = [3,1,0]$, the following inequality follows immediately from the Muirhead's inequality (6.5)

$$2!\left(x_1^4 + x_2^4 + x_3^4\right) \geq x_1^3x_2 + x_1^3x_3 + x_2^3x_1 + x_2^3x_3 + x_3^3x_1 + x_3^3x_2 \tag{6.10}$$

which is effectively inequality (6.5).

Inequalities (6.1)–(6.4) can be interpreted in a natural way if the constraints $0 \leq x_1 \leq 1$, $0 \leq x_2 \leq 1$ and $0 \leq x_3 \leq 1$ are imposed on the variables entering the inequalities. If the left- and right-hand side of inequality (6.1) are multiplied by 1/3, the inequality

$$(1/3)x_1^2 + (1/3)x_2^2 + (1/3)x_3^2 \geq (1/3)x_1x_2 + (1/3)x_2x_3 + (1/3)x_3x_1 \tag{6.11}$$

is obtained, whose left- and right-hand side can be meaningfully interpreted by using probabilities of mutually exclusive events. The factors 1/3 on the left-hand side of inequality (6.11) can be interpreted as probabilities of random selection of a particular supplier from three available suppliers (Figure 6.1). Each supplier offers

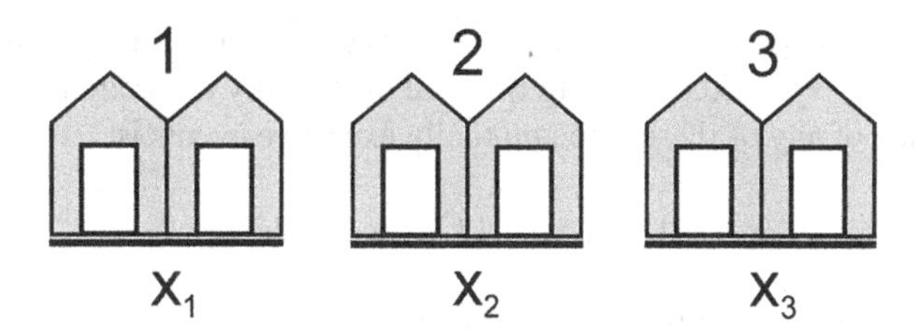

FIGURE 6.1 Three suppliers supplying high-reliability components with unknown proportions x_1, x_2 and x_3.

components of the same type and the variables x_1, x_2 and x_3 can be interpreted as the fractions of high-reliability components characterising the individual suppliers $(0 \le x_i \le 1)$, correspondingly. It is important to emphasise that the fractions x_i of high-reliability components characterising the individual suppliers *are unknown.*

In the case of suppliers of suspension automotive springs, for example, high-reliability components mean that only a fraction x_i, where $i = 1,2,3$, of the manufactured suspension springs can last for more than a specified number of cycles if tested on a specially designed test rig and the rest of the springs fail significantly below the specified number.

The left-hand side of inequality (6.11) can be interpreted as the probability of purchasing two high-reliability components from a randomly selected supplier.

Indeed, purchasing two high-reliability components from the same supplier can occur in three mutually exclusive ways: (i) the first supplier is randomly selected and both components purchased from the first supplier are highly reliable, the probability of which is $(1/3)x_1^2$; (ii) the second supplier is randomly selected and both components purchased from the second supplier are highly reliable, the probability of which is $(1/3)x_2^2$ and finally, (iii) the third supplier is randomly selected and both components purchased from the third supplier are highly reliable, the probability of which is $(1/3)x_3^2$. According to the total probability theorem, the probability of occurrence of any of these three mutually exclusive events is

$$p_1 = (1/3)x_1^2 + (1/3)x_2^2 + (1/3)x_3^2$$

which is the left-hand side of inequality (6.11). It is assumed that the amount of components offered by each supplier is sufficiently large so that the probability of purchasing a second high-reliability component is equal to the probability that the first purchased component will be of high reliability.

The right-hand side of inequality (6.11) can be interpreted as the probability of purchasing two high-reliability components from two different, randomly selected suppliers. Indeed, purchasing two high-reliability components from two suppliers can occur in three mutually exclusive ways: (i) suppliers 1 and 2 are randomly selected, and both components purchased from these suppliers are highly reliable (with probability $(1/3)x_1x_2$); (ii) suppliers 2 and 3 are randomly selected, and both components purchased from these suppliers are highly reliable (with probability $(1/3)x_2x_3$) and finally, (iii) suppliers 3 and 1 are randomly selected, and the components purchased from these suppliers are highly reliable (with probability $(1/3)x_3x_1$). According to the total probability theorem, the probability of occurrence of any of these three mutually exclusive events is

$$p_2 = (1/3)x_1x_2 + (1/3)x_2x_3 + (1/3)x_3x_1$$

which is the right-hand side of inequality (6.11).

The prediction of inequality (6.11) states that purchasing both components from a single, randomly selected supplier is characterised by a higher probability of purchasing components that are both of high reliability. This is *a surprising and highly counter-intuitive result*. After all, the proportions x_i of high-reliability components characterising the suppliers are unknown.

The difference in the probabilities p_1 and p_2 evaluated from the left- and right-hand side of inequality (6.11) can be significant. Thus, for $x_1 = 0.5$, $x_2 = 0.9$ and $x_3 = 0.1$, the left-hand side of inequality (6.11) gives

$$p_1 = (1/3) \times 0.5^2 + (1/3) \times 0.9^2 + (1/3) \times 0.1^2 = 0.357,$$

while the right-hand side gives

$$p_2 = (1/3) \times 0.5 \times 0.9 + (1/3) \times 0.9 \times 0.1 + (1/3) \times 0.1 \times 0.5 = 0.197$$

Equality in inequality (6.11) is attained only for $x_1 = x_2 = x_3$.

Similar interpretation can be made for inequality (6.2). If the left- and right-hand side of inequality (6.2) are multiplied by $1/3$, the inequality

$$(1/3)x_1^3 + (1/3)x_2^3 + (1/3)x_3^3 \geq x_1x_2x_3 \tag{6.12}$$

is obtained.

The left-hand side of inequality (6.12) is the probability of purchasing three high-reliability components from a randomly selected supplier. The right-hand side of inequality (6.12) is the probability of purchasing three high-reliability components from the three available suppliers.

Again, inequality (6.12) predicts that if the proportions x_i of high-reliability components characterising the suppliers are unknown, purchasing the three components from a single, randomly selected supplier is associated with higher probability that all purchased components will be of high reliability.

Next, consider inequality (6.3). If the left- and right-hand side of inequality (6.3) are multiplied by $1/6$, the inequality

$$\begin{aligned}(1/3)x_1^3 + (1/3)x_2^3 + (1/3)x_3^3 \geq (1/6)x_1^2x_2 + (1/6)x_1^2x_3 + (1/6)x_2^2x_1 \\ + (1/6)x_2^2x_3 + (1/6)x_3^2x_1 + (1/6)x_3^2x_2\end{aligned} \tag{6.13}$$

is obtained.

The left-hand side of inequality (6.13) is the probability of purchasing three high-reliability components from a randomly selected supplier. The right-hand side of inequality (6.13) is the probability of purchasing three high-reliability components from two randomly selected suppliers. Again, inequality (6.13) predicts that purchasing all components from a single, randomly selected supplier is associated with a higher probability of purchasing three high-reliability components.

Finally, consider inequality (6.4). If the left- and right-hand side of inequality (6.4) are multiplied by 1/6, the inequality

$$(1/3)x_1^4 + (1/3)x_2^4 + (1/3)x_3^4 \geq (1/6)x_1^3 x_2 + (1/6)x_1^3 x_3 + (1/6)x_2^3 x_1 + (1/6)x_2^3 x_3 + (1/6)x_3^3 x_1 + (1/6)x_3^3 x_2 \qquad (6.14)$$

is obtained.

The left-hand side of inequality (6.14) is the probability of purchasing four high-reliability components from a randomly selected supplier. The right-hand side of inequality (6.14) is the probability of purchasing four high-reliability components from two suppliers by purchasing three components from one randomly selected supplier and the other component from another randomly selected supplier.

In summary, if no information is available about the fractions of high-reliability components characterising the individual suppliers, the best strategy of purchasing only high-reliability components is to purchase the components from a single, randomly selected supplier. Despite the complete lack of knowledge related to the proportions of high-reliability components characterising the separate suppliers and despite existing dependencies among the suppliers, purchasing all components from the same supplier is characterised by the highest probability that all purchased components will be of high reliability.

These highly counter-intuitive results fly in the face of the conventional habit advocating diversification as a risk reduction measure and expose the dangers of blindly following conventional wisdom in risk reduction.

6.2 INCREASING THE PROBABILITY OF SUCCESSFUL ACCOMPLISHMENT OF TASKS BY DEVICES WITH UNKNOWN RELIABILITY

The inequalities from the previous section have alternative interpretations. Consider, for example, inequalities (6.12) and (6.13), which, for convenience, will be presented in the form:

$$\frac{1}{3}x^3 + \frac{1}{3}y^3 + \frac{1}{3}z^3 \geq xyz \qquad (6.15)$$

$$(1/3)x^3 + (1/3)y^3 + (1/3)z^3 \geq (1/6)x^2 y + (1/6)x^2 z + (1/6)y^2 x + (1/6)y^2 z + (1/6)z^2 x + (1/6)z^2 y \qquad (6.16)$$

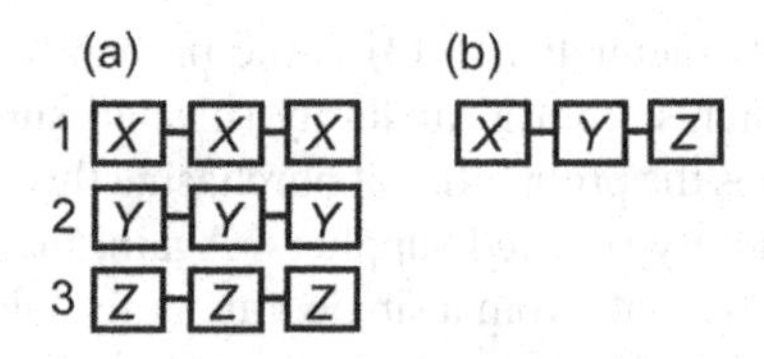

FIGURE 6.2 (a) An arrangement of three devices of the same type allocated to three identical tasks; (b) an arrangement of three devices of different types assigned to three identical tasks.

The factors 1/3 on the left-hand side of inequality (6.15) can be interpreted as probabilities of selecting any of the three arrangements (Figure 6.2a; arrangements 1, 2 and 3). Each arrangement includes three devices of types *X*, *Y* and *Z* assigned to a mission including three identical tasks. The probabilities x, y and z of successful accomplishment of a task by a device of type *X*, *Y* and *Z* *are unknown.*

A mission is considered to have been accomplished successfully if all three tasks assigned to the devices have been accomplished successfully.

According to the total probability theorem, the left-hand side of inequality (6.15) represents the total probability that a randomly selected arrangement consisting of three devices of the same type, allocated to three identical tasks (Figure 6.2a), will accomplish successfully a mission.

Indeed, a successful accomplishment of a mission by devices of the same type can occur in three different, mutually exclusive ways: (i) arrangement 1 including devices of type *X* is randomly selected, and all devices successfully accomplish their tasks, the probability of which is $(1/3)x^3$; (ii) arrangement 2 including devices of type *Y* is randomly selected, and all devices successfully accomplish their tasks, the probability of which is $(1/3)y^3$ and finally, (iii) arrangement 3 including devices of type *Z* is randomly selected, and all devices successfully accomplish their tasks, the probability of which is $(1/3)z^3$.

The right-hand side of inequality (6.15) is the probability xyz that the devices of different types (*X*, *Y* and *Z*) successfully accomplish their tasks.

The separate terms of inequality (6.16) can also be meaningfully interpreted by using the total probability theorem for mutually exclusive events. According to the total probability theorem, the left-hand side of inequality (6.16) represents the total probability that a randomly selected arrangement consisting of three identical type devices will successfully accomplish the mission (Figure 6.3a).

According to the total probability theorem, the right-hand side of inequality (6.16) represents the total probability that a randomly selected arrangement composed of three devices, two of which are of the same type, will successfully accomplish the mission. A successful accomplishment of a mission including three devices, two of which are of the same type, can occur in six distinct ways (Figure 6.3b): (i) the device arrangement (X, X, Y) is randomly selected, and all three tasks are successfully accomplished by the devices, the probability of which is $(1/6)x^2y$; (ii) the device arrangement (X, X, Z) is randomly selected, and all three tasks are successfully accomplished by the devices, the probability of which is $(1/6)x^2z$ and so on.

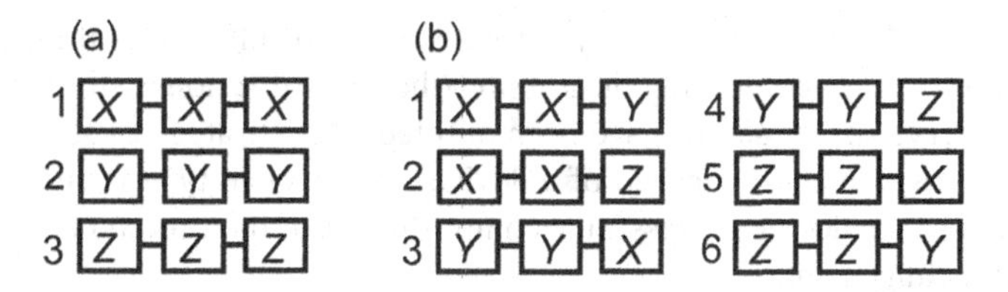

FIGURE 6.3 (a) Arrangements of three devices of the same type assigned to three identical tasks; (b) arrangements of three devices, two of which are of the same type, assigned to three identical tasks.

Despite that the probabilities x, y and z of successful accomplishment of a task by the devices of different types are *unknown*, according to the predictions of inequalities (6.15) and (6.16), the strategy of randomly selecting an arrangement composed of three devices of the same type is characterised by a higher chance of a successful accomplishment of the mission compared to selecting an arrangement composed of three different types of devices or an arrangement composed of three devices, two of which are of the same type.

This is a counter-intuitive result considering that the probabilities x, y and z of successful accomplishment of a task by the different types of devices are *unknown*. Why should any conceivable strategy give an edge if the probabilities x, y and z are unknown? It seems that in such a situation of deep uncertainty, no particular strategy matters, yet this impression is wrong. Despite the existing deep uncertainty related to the probabilities of successful accomplishment of the tasks by the devices, selecting randomly an arrangement composed of the same type devices is always the best strategy. This conclusion holds *irrespective of existing unknown interdependencies among the probabilities of successful accomplishment of the tasks from the devices.*

The significant advantage provided by the superior strategy can be illustrated by a numerical example. Suppose that the probabilities of accomplishing the tasks characterising the different device types X, Y and Z are $x = 0.88$, $y = 0.64$ and $z = 0.38$. According to inequality (6.15), the probability that a randomly selected arrangement including three devices of the same type will successfully accomplish the mission is

$$p_1 = (1/3) \times 0.88^3 + (1/3) \times 0.64^3 + (1/3) \times 0.38^3 = 0.33,$$

while the probability of accomplishing successfully the mission by three devices of different types is $p_2 = 0.88 \times 0.64 \times 0.38 = 0.21$.

According to inequality (6.16), the probability of accomplishing the mission by a randomly selected arrangement consisting of three devices, two of which are of the same type, is

$$p_3 = (1/6) \times [0.88^2 \times 0.64 + 0.88^2 \times 0.38 + 0.64^2 \times 0.88 + 0.64^2 \times 0.38 \\ + 0.38^2 \times 0.88 + 0.38^2 \times 0.64] = 0.25$$

By using the Muirhead's inequality for $n>3$, these results are naturally generalised for $n>3$ tasks composing the mission (the generalisation is straightforward and to conserve space, details have been omitted). The strategy of selecting randomly an arrangement composed of n devices of the same type is characterised by the highest chance of successful accomplishment of the mission compared to any other arrangement.

6.3 MONTE CARLO SIMULATIONS

The predictions from inequalities (6.11)–(6.14) have been confirmed by Monte Carlo simulations, each of which involved 10 million trials. For fractions of high-reliability components of 0.9, 0.55 and 0.35, characterising the three suppliers, the Monte Carlo simulation resulted in probabilities $p_1 = 0.41$ and $p_2 = 0.33$ of purchasing two high-reliability components from a randomly selected single supplier and from two randomly selected suppliers, correspondingly. The left- and right-hand side of inequality (6.11) give $p_1 = 0.41$ and $p_2 = 0.33$ for the same probabilities, which confirms the validity of inequality (6.11).

For fractions of high-reliability components characterising the three suppliers, $x_1 = 0.9$, $x_2 = 0.75$ and $x_3 = 0.25$, the Monte Carlo simulation based on 10 million trials resulted in probabilities $p_1 = 0.389$, $p_2 = 0.26$ and $p_3 = 0.169$ of purchasing three high-reliability components from a randomly selected single supplier, from two randomly selected suppliers and from all three available suppliers, correspondingly. The left- and right-hand sides of inequalities (6.15) and (6.16) yield $p_1 = 0.389$, $p_2 = 0.26$ and $p_3 = 0.169$ for the same probabilities, which confirms the validity of inequalities (6.15) and (6.16).

For high-reliability fractions of 0.9, 0.4 and 0.3 characterising the suppliers, the Monte Carlo simulations yielded the value 0.23 for the probability of four high-reliability components purchased from a randomly selected supplier and the value 0.1 for the probability of four high-reliability components if three components are purchased from a randomly selected supplier and the remaining component from another randomly selected supplier. These results are confirmed by the probabilities calculated from the left- and right-hand side of inequality (6.14).

6.4 ASSESSING THE EXPECTED TIME OF UNSATISFIED DEMAND FROM USERS PLACING RANDOM DEMANDS ON A TIME INTERVAL

Complex algebraic inequalities can also be given meaningful interpretation. Consider the complex algebraic inequality

$$(1-\psi)^n + n\psi^1(1-\psi)^{n-1} + \frac{n(n-1)}{1\times 2}\psi^2(1-\psi)^{n-2} + \cdots + \frac{n(n-1)\ldots(n-m)}{1\times 2\times\ldots\times m}\psi^m(1-\psi)^{n-m} \le 1 \tag{6.17}$$

where $0 \le \psi \le 1$, n and m are integers for which $m \le n$; $n > 1$.

Inequality (6.17) can be proved easily by observing that it has been obtained from the binomial expansion of the expression

$$[(1-\psi)+\psi]^n = 1 \tag{6.18}$$

and discarding from the expansion all positive terms corresponding to $m+1, m+2, \ldots, n$. Equality in inequality (6.17) is attained for $\psi = 0$ or $m = n$.

Inequality (6.17) can be interpreted if n stands for the number of users placing random demands for a particular resource over the time interval $(0,L)$ and ψ stands for the time fraction of random demand over the time interval $(0,L)$. For example, if the duration of each random demand is d, $\psi = d / L$ is the time fraction of random demand.

Consider users demanding randomly a particular single resource (piece of equipment, repairers, process, etc.), during an operation period with length L. The available single resource (e.g. piece of equipment) can satisfy only a single random demand at a time. The random appearance of demand i is marked by s_i and the end of demand i is marked by e_i. Figure 6.4 depicts three random demands with durations of length d appearing at random times s_1, s_2 and s_3. The overlapping region $s_3\ e_2$ in the figure marks the simultaneous appearance of two random demands which leads to unsatisfied demand.

For complex systems (production system, computer network, etc.), the random demands can be demands for repair from the failed components building the system. The random demands can also be demands for a particular piece of life-saving equipment from critically ill people, etc.

The first term $(1-\psi)^n$ of inequality (6.17) is the expected fraction of time during which no random demand is present. Indeed, the expected fraction of time during which no random demand is present is equal to the probability that a randomly selected point in the time interval $(0,L)$ will not be 'covered' by any of the n randomly placed demands (segments with length d).

The second term $n\psi^1(1-\psi)^{n-1}$ of inequality (6.17) is the expected fraction of time during which exactly one random demand is present in the time interval $(0,L)$. Indeed, this expected time fraction is equal to the probability that a randomly selected point in the time interval $(0,L)$ will be covered by exactly one of the n randomly placed demands (segments with length d). This is a sum of the probabilities $\psi^1(1-\psi)^{n-1}$ of n mutually exclusive events: exactly one segment of length d covers the selected point and none of the other segments covers it.

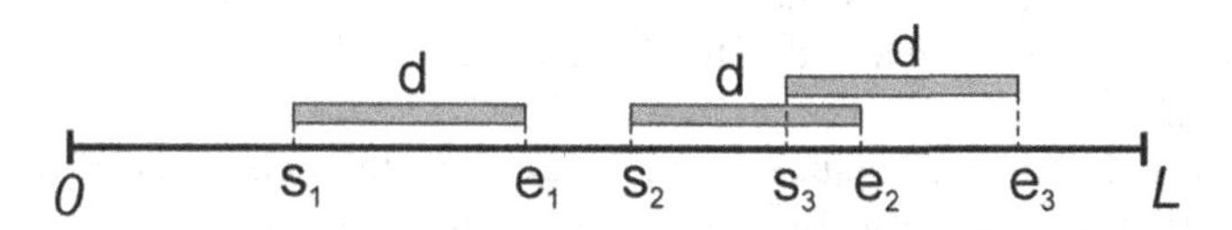

FIGURE 6.4 Random demands for a particular resource over a time interval 0,*L*.

The third term $\frac{n(n-1)}{1 \times 2}\psi^2(1-\psi)^{n-2}$ of inequality (6.17) is the expected fraction of time during which exactly two random demands are present. This time fraction is equal to the probability that a randomly selected point in the time interval $(0,L)$ will be covered by exactly two of the n randomly placed segments with length d. This is a sum of the probabilities $\psi^2(1-\psi)^{n-2}$ of $\frac{n(n-1)}{1 \times 2}$ mutually exclusive events: exactly two segments of length d cover the selected point and none of the other segments covers it.

The $m+1st$ term $\frac{n(n-1)\ldots(n-m+1)}{1 \times 2 \times \ldots \times m}\psi^m(1-\psi)^{n-m}$ of inequality (6.17) is the expected fraction of time during which exactly m random demands are present. This time fraction is equal to the probability that a randomly selected point in the time interval $(0,L)$ will be covered by exactly m of the n randomly placed segments with length d. It is a sum of the probabilities $\psi^m(1-\psi)^{n-m}$ of $\frac{n(n-1)\ldots(n-m+1)}{1 \times 2 \times \ldots \times m}$ mutually exclusive events: exactly m segments representing the random demands from m users cover the selected point and none of the other segments covers it.

The sum of the expected fractions of time representing $0,1,2,\ldots, n$ simultaneous random demands is equal to 1 because $[(1-\psi)+\psi]^n = 1$.

Because the sum of all expected time fractions corresponding to $0,1,2,\ldots, n$ simultaneous random demands is equal to 1, inequality (6.17) effectively states that the sum of the expected fractions of time during which there is no random demand, there is exactly one random demand, exactly two random demands,..., exactly m random demands, cannot exceed unity.

Suppose that the available resources can satisfy m (or fewer) simultaneous demands, but not more than m simultaneous demands. Because the sum of all expected time fractions corresponding to $0,1,2,\ldots, n$ simultaneous random demands is equal to 1, the total expected fraction of time during which there is unsatisfied demand ($m+1$ or more random demands are simultaneously present) is given by

$$p = 1 - \left((1-\psi)^n + n(1-\psi)^{n-1}\psi^1 + \frac{n(n-1)}{1 \times 2}(1-\psi)^{n-2}\psi^2 \right.$$
$$\left. + \cdots + \frac{n(n-1)\ldots(n-m)}{1 \times 2 \times \ldots \times m}(1-\psi)^{n-m}\psi^m \right) \geq 0$$

This inequality is effectively the rearranged inequality (6.17) and p is the probability of unsatisfied demand due to clustering of random demands in the time interval $(0,L)$.

Clearly, the compact inequality (6.17) describes the behaviour of a complex system with a number of features: (i) an arbitrary number n of users initiating demands, (ii) each demand is randomly placed along a specified time interval, (iii) different number of sources available for servicing the random demands.

It can be shown that inequality (6.17) is valid also if the demand time is not fixed but varies, with mean equal to d (Todinov, 2017).

Inequality (6.17) has a wide range of applications.

- The competition of random demands for a particular single resource/service on a finite time interval is a common example of risk controlled by the simultaneous presence of critical events. The appearance of a critical event engages the servicing resource and if a new critical event occurs during the service time of the first critical event, no servicing resource will be available for the second event.

Suppose that only a single repair unit is available for servicing failures on a power line. If a power line failure occurs, the repair resource will be engaged and if another failure occurs during the repair time d associated with the first failure, no free repair resource will be available to recover from the second failure. The delay associated with the second repair could lead to overloading of the power distribution system, thereby inducing further failures.

There are cases where the probability of a simultaneous presence (overlapping) of risk-critical critical events must be low. A low probability of a simultaneous presence of random demands is required in situations where people in critical condition demand a particular piece of life-saving equipment for time interval with length d. If only a single piece of life-saving equipment is available, simultaneous demands within a time interval with length d cannot be satisfied and the consequences could be fatal.

- Stored spare equipment in a warehouse servicing the needs of customers arriving randomly during a specified time interval. After a demand from a customer, the warehouse needs a minimum period of length d to restore/return the dispatched equipment before the next demand can be serviced. In this case, the probability of unsatisfied demand equals the probability of clustering of two or more customer arrivals within the critical period needed for making the equipment available for the next customer.
- Supply systems that accumulate the supplied resource before it is dispatched for consumption (compressed gaseous substances, for example). Suppose that, after a demand for the resource, the system needs a minimum period of length d to restore the amount of supplied resource to the level existing before the demand. In this case, the probability of unsatisfied demand equals the probability of clustering of two or more random demands within the critical recovery period with length d.
- Risk controlled by overlapping (simultaneous presence) of critical events is present when the appearance of one critical event (e.g. a shock) requires a particular time during which the system needs to recover. If another critical event appears before the system has recovered, the system's strength/capacity is exceeded which results in system failure.

Consider, for example, shocks caused by failures associated with pollution to the environment (e.g. a leakage of chemicals). A failure, followed by another failure associated with leakage of chemicals before a critical recovery time interval has elapsed, could result in irreparable damage to the environment. For example, clustering of failures associated with a release of chemicals in the sea water could result in a dangerously high acidity which will destroy marine life.

In all of the listed examples, the overlapping of the risk-critical random events cannot be avoided; therefore, the level of risk correlates with the expected time fraction of overlapping risk-critical events.

7 Enhancing Decision-Making by Interpretation of Algebraic Inequalities

7.1 MEANINGFUL INTERPRETATION OF AN ALGEBRAIC INEQUALITY RELATED TO RANKING THE MAGNITUDES OF SEQUENTIAL RANDOM EVENTS

Algebraic inequalities do not need to be complex, in order to extract from them useful knowledge by interpretation. The next example, involving the interpretation of a very simple algebraic inequality with far-reaching consequences, has been prompted by the short note by Cover (1987). Consider the simple algebraic inequality (Todinov, 2020b)

$$p^2 + (1-p)^2 + 4p(1-p) \geq 1 \tag{7.1}$$

where $0 \leq p \leq 1$. This inequality can be proved easily by simplifying the left-hand side to $1 + 2p(1-p)$ which is obviously not smaller than 1 because $p(1-p) \geq 0$. If both sides of inequality (7.1) are divided by 2, inequality (7.1) can be transformed into the next interpretable, equivalent inequality:

$$(1/2)p^2 + (1/2)(1-p)(1-p) + (1-p)p + p(1-p) \geq 1/2 \tag{7.2}$$

Inequality (7.2) can be interpreted by creating meaning for the variable p and for the terms in the left-hand side of the inequality. Because $0 \leq p \leq 1$, it is natural to interpret the variable p as 'probability'. If p is the probability that the magnitude y of a random event with a uniform distribution exceeds a specified threshold C (Figure 7.1), the terms p^2, $(1-p)^2$, $(1-p)p$ and $p(1-p)$ can be defined as follows: p^2 is the probability that the magnitudes of both statistically independent random events exceed the fixed threshold C (Figure 7.1a); $(1-p)^2$ is the probability that the magnitudes of both random events are below the threshold C (Figure 7.1b); $(1-p)p$ is the probability that the magnitude of the first random event is below the threshold C and the magnitude of the second random event is above the threshold C (Figure 7.1c) and $p(1-p)$ is the probability that the magnitude of the first

DOI: 10.1201/9781003199830-7

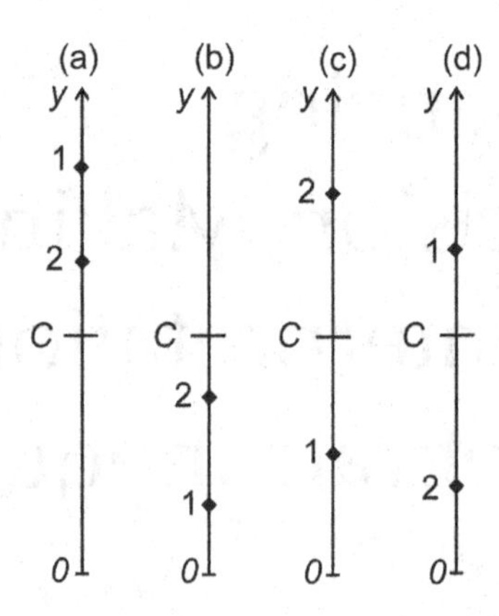

FIGURE 7.1 Possible locations of the magnitudes of two random events with respect to a threshold C: (a) both magnitudes above the pre-selected threshold C; (b) both magnitudes below the pre-selected threshold C; first event magnitude below the threshold C, second event magnitude above the threshold C; (d) first event magnitude above the threshold C, second event magnitude below the threshold C.

random event is above the threshold C and the magnitude of the second random event is below the threshold C (Figure 7.1d).

The first term $(1/2)p^2$ in inequality (7.2) can then be interpreted as the probability that the magnitudes of two sequential random events will both exceed the selected threshold C and the second random event will have a smaller magnitude than the first random event (Figure 7.1a).

The second term $\frac{1}{2}(1-p)(1-p)$ can be interpreted as the probability that the magnitudes of two sequential random events will both be below the selected threshold C and the second random event will have a larger magnitude than the first random event (Figure 7.1b).

The third term $1 \times (1-p)p$ can be interpreted as the probability that the first random event will be below the threshold C, the second random event will be above the threshold C and will have a larger magnitude than the first random event (Figure 7.1c).

Finally, the term $1 \times p(1-p)$ can be interpreted as the probability that the first random event will be above the threshold C, the second random event will be below the threshold C and will have a smaller magnitude than the first random event (Figure 7.1d).

Consider now a real-world problem related to predicting magnitude ranking of safety-critical random events, important in cases where it is necessary to decide on whether there is a need for extra resources to mitigate the consequences associated with future random events. Such random events could be, for example, the leaks of crude oil into marine water from a pipeline during subsea oil and gas production, the magnitudes of floods in consecutive years, the magnitudes of fires and the direction of movement of the price of a commodity on the stock market. If no information is available about the distribution of the magnitude of the random event, it seems that the chance of guessing correctly whether the magnitude of the second random event will be larger than the magnitude of the first random event is only 50% and cannot be improved.

According to a prediction strategy outlined in Cover (1987), if the magnitude of the first random event is larger than the randomly pre-selected comparison threshold C (Figure 7.1a), a prediction is made that the second random event will have a smaller magnitude than the magnitude of the first random event. If the magnitude of the first random event is smaller than the comparison threshold C, a prediction is made that the second random event will have a larger magnitude than the magnitude of the first random event (Figure 7.1b).

According to inequality (7.2), by following this prediction strategy, the probability of predicting correctly the ranking of the magnitudes of the random events can be greater than 50%.

Indeed, in the cases where both random events have larger magnitudes than the pre-selected threshold C, or both random events have smaller magnitudes, the likelihood of a correct prediction related to the magnitude of the second random event is clearly 0.5. In the case where the first random event has a magnitude smaller than the threshold C and the second random event has a larger magnitude, the likelihood of a correct prediction related to the magnitude of the second random event is 1.0. Similarly, in the case where the first random event has a larger magnitude than the pre-selected threshold and the second event has a smaller magnitude, the likelihood of a correct prediction related to the magnitude of the second random event is also 1.0.

Contrary to the statement made in Cover (1987), however, from inequality (7.2), it follows that the prediction strategy described earlier *does not ensure correct prediction with probability strictly greater than 0.5.*

Indeed, if the threshold C had been selected to be too large or too small within the physically possible range of values for the magnitude of the random event, then the selected threshold would underestimate or overestimate the expected magnitude and $1 - p \approx 0$ or $p \approx 0$. In this case, the quantity $p(1 - p) \approx 0$ on the left-hand side of inequality (7.2) is very small and, practically, a probability of correct prediction beyond 0.5 cannot be attained.

The product of two positive values with a given sum has a maximum when the values are equal. Consequently, the product pq where $p + q = 1$ attains maximum when $p = q = 1/2$. As a result, the maximum possible value of the quantity $p(1 - p)$ is 0.25 and this value cannot be exceeded. Consequently, the most appropriate position of the threshold C is at a value for which the probability that the magnitude of the event will be smaller than the threshold is equal to the probability that the magnitude of the event will be larger than the threshold. In this case, correct prediction will be made in 75% of the trials.

In reality, the threshold position of equal probability is not known and the threshold C is selected randomly within the physically possible range of values for the magnitude of random events.

To test the described decision strategy, a Monte Carlo simulation experiment has been made (Todinov, 2020b). Suppose that the leak of crude oil is from a pipeline with a maximum debit of 300 *l/s*. Because no leak can be negative and no oil leak can exceed the maximum debit of the pipeline, the lower and upper limit of the range of all possible leaks are 0 and 300, correspondingly. The possible leak magnitudes within the physically possible range are uncertain, this is why the distribution of the leak magnitude is assumed to be uniform, with randomly selected lower and upper limit over the possible physical range [0,300].

Series of Monte Carlo simulations each consisting of 10 million random trials resulted in probabilities of a correct prediction within the range [0.50–0.64], which confirmed that the proposed strategy, indeed, reduces uncertainty and yields a prediction result related to the magnitude ranking of the second leak which is better than a random guess. If the prediction is made randomly, without a comparison with a pre-selected threshold, the probability of a correct prediction is exactly 50%.

If a new meaning is attached to the event, the abstract inequality (7.2) can be applied to other real processes: stock price, magnitudes of random shocks on a device or a structure, etc. The described interpretation can also be used to transform a neutral bet with expected profit zero into a good bet with positive expected profit. A large number of good bets materialises into gain (Todinov, 2013).

7.2 IMPROVING PRODUCT RELIABILITY BY INCREASING THE LEVEL OF BALANCING

Well-balanced systems and assemblies distribute loads more uniformly across components and exhibit higher reliability.

Improving the level of balancing in systems and assemblies can be achieved by

- ensuring more uniform load distribution among components,
- ensuring conditions for self-balancing and
- reducing the variability of risk-critical parameters

7.2.1 Ensuring More Uniform Load Distribution among Components

Ensuring a more uniform distribution of the load is essential for both electrical and mechanical components. In mechanical assemblies, for example, where torque is transmitted from a rotating shaft to a gear, sprocket of pulley or *vice versa,* using splines instead of keys improves the load distribution, decreases the stresses and ensures higher resistance of the connection to failure in the case of overload. In another example, a flange connection with a very small number of fasteners leads to excessive stresses in some of the fasteners. Increasing the number of fasteners improves the load distribution and reduces the risk of failure.

Segmentation of a load-carrying component often results in increased contact area and better conforming contact which reduces contact stresses and improves reliability.

Improved load distribution achieved by segmentation, increases the level of balancing in rotating components which results in greater stability, smaller vibration amplitudes, smaller inertia forces and enhanced reliability. Examples of improved load distribution are the introduction of multiple cylinders in an internal combustion engine, multiple blades on a turbine, etc.

7.2.2 Ensuring Conditions for Self-Balancing

Unbalanced forces cause premature wear-out, fatigue degradation and failure. As a rule, improving the level of balancing in a system reduces the magnitudes of these forces, the loading stresses and improves system reliability.

Self-balancing is present in cases where negative factors are made to cancel one another. The result is improved capability of the system to adapt to adversity and recover which enhances the resilience of the system.

Self-balancing is frequently achieved through symmetrical design which eliminates unwanted forces and moments in rotating machinery. Thus, self-balancing through symmetrical design can be used to minimise the axial forces on turbine shafts (Matthews, 1998). Thus, the axial forces in herringbone gear meshing are counterbalanced, which eliminates the need for thrust bearings. An increase in the transmitted torque increases simultaneously the magnitude of each axial force and, because they act in opposite directions, the result is a very small resultant axial force. Twisting wires to cancel their magnetic interference is also an example of self-balancing. Twisted wires carry equal and opposite currents whose electromagnetic fields cancel.

7.2.3 Reducing the Variability of Risk-Critical Parameters

Reducing the variability of risk-critical parameters prevents them from reaching dangerous levels. Reliability and risk-critical parameters vary and this variability can be broadly divided into the following categories: (i) variability associated with material and physical properties, manufacturing and assembly; (ii) variability caused by product deterioration; (iii) variability associated with the loads the product experiences in service and (iv) variability associated with the operating environment.

Strength variability caused by production variability and variability of properties is one of the major reasons for an increased interference of the strength distribution and the load distribution which results frequently in overstress failures (Todinov, 2016). A heavy lower tail of the distribution of properties usually yields a heavy lower tail of the strength distribution, thereby promoting early-life failures.

Selecting components of the same variety is an important method for reducing variability and improving the level of balancing.

The requirement for selecting components of the same variety ensures a more uniform distribution of the load which is essential for both electrical and mechanical components. This requirement, for example, is essential for transistors in power supply circuits, mechanical components in contact, resistors in sensitive bridge circuits, bearings in rotating shafts and belts in belt drives. Thus, selecting transistors of the same variety working in parallel in a power supply circuit is an important measure for ensuring a uniform distribution of the load on the transistors and avoiding overheating and premature failure of any of the transistors.

For mechanical components in contact, selecting both components of the same variety (for example, both with high surface hardness or both with normal surface hardness) is an important measure to reduce the rate of wear and avoid premature wear-out of one of the components.

Suppose that a particular key property (tolerance, strength, weight, etc.) is obtained from pooling n batches, each containing m_i components. In each batch, the key property follows a particular unknown distribution with mean μ_i and standard deviation σ_i, where $i = 1, \ldots, n$. If all batches have been pooled together in a single, large batch, the distribution of the key property in the single, large batch is a mixture of n distributions, where $p_i = \frac{m_i}{\sum_{k=1}^{n} m_k}$ is the probability of selecting a component from the ith batch. The mean μ of the distribution of the key property in the pooled batch is given by

$$\mu = \sum_{i=1}^{n} p_i \mu_i, \tag{7.3}$$

while the variance $V = \sigma^2$ of the property in the pooled batch is given by (Todinov, 2002b, 2003)

$$V = \sum_{i=1}^{n} p_i [\sigma_i^2 + (\mu_i - \mu)^2] \tag{7.4}$$

As can be seen, the variance of the property in the pooled batch can be decomposed into two major components. The first component $\sum_{i=1}^{n} p_i \sigma_i^2$ characterises only the variation of properties within the batches while the second component $\sum_{i=1}^{n} p_i (\mu_i - \mu)^2$ of equation 7.4 characterises the variation of properties between the separate batches. Assuming that all individual distributions have the same mean μ ($\mu_i = \mu_j = \mu$), the terms $p_i (\mu_i - \mu)^2$ in equation 7.4, related to between-sources variation, become zero and the total variance becomes $V = \sum_{i=1}^{n} p_i \sigma_i^2$. In other words, in this case, the total variation of the property is entirely a within-sources variation (Todinov, 2003).

The variation of properties can be reduced significantly if all components are selected from batch k, characterised by the smallest variance σ_k^2. In this case, selecting components from the same batch (the same variety of components) reduces the variance from $V = \sum_{i=1}^{n} p_i [\sigma_i^2 + (\mu_i - \mu)^2]$ to $V = \sigma_k^2$.

Selecting components of the same variety is a key measure in reducing variability of properties and improving the level of balancing. In many cases, the probability of selecting items of the same variety can be equated to the probability of building a reliable assembly.

7.3 ASSESSING THE PROBABILITY OF SELECTING ITEMS OF THE SAME VARIETY TO IMPROVE THE LEVEL OF BALANCING

Consider the common abstract inequality

$$a^2 + b^2 \geq 2ab \quad (7.5)$$

which is true for any real numbers a and b because the inequality can be obtained directly from the obvious inequality $(a-b)^2 \geq 0$. Suppose that the numbers a and b are not both equal to zero. If both sides of inequality (7.5) are divided by the positive quantity $(a+b)^2$, the next inequality

$$\frac{a^2}{(a+b)^2} + \frac{b^2}{(a+b)^2} \geq \frac{2ab}{(a+b)^2} \quad (7.6)$$

is obtained, whose parts have a very useful interpretation (Todinov, 2020c). If a and b are the number of items of variety A and variety B in a large batch of items, the left-hand side of inequality (7.6) is the probability that from two randomly selected items from the batch, both will be of the same variety (variety A only or variety B only).

The varieties can be, for example, components manufactured by two different machine centres. It is assumed that the batch is sufficiently large so that if two items are taken from the batch, the probability that the second item will be of a particular variety is practically independent of the variety of the first item.

The right-hand side of inequality (7.6) can be interpreted as the probability that the two randomly selected items will be of different variety (A, B or B, A). The two selected components can be either of the same variety or of different variety. Therefore, the sum of the probabilities of the events 'the components are of the same variety' and 'the components are of different variety' add up to unity. Consequently, inequality (7.6) predicts that the probability that the two selected items will be of different variety is smaller than 0.5 or at most equal to 0.5 but never exceeds 0.5. This prediction is counter-intuitive, considering the symmetry of the cases leading to the same-variety outcome and to a different-variety outcome. Because of symmetry, for example, flipping two identical coins is equally likely to result in the same outcome on both coins (head/head or tail/tail) or in different outcomes on the coins (head/tail or tail/head).

The difference between the probabilities calculated from the left- and right-hand side of inequality (7.6) can be very large. Thus, for $a = 300$ items of variety A and 800 items of variety B, the left-hand side of inequality (7.6) yields probability of 0.603 for selecting two components of the same variety and probability of 0.397 that the selected components will be of different variety.

Interestingly, no analogous inequality exists for a larger than two number of randomly selected components from a large batch containing two different variety. Thus, the probability of selecting n components of the same variety ($n > 2$) from a large batch ($n \ll a + b$) is given by

$$p_1 = a^n/(a+b)^n + b^n/(a+b)^n, \tag{7.7}$$

while the probability of not selecting n components of the same variety is $p_2 = 1 - p_1$ as probabilities of complementary events.

Denoting $u = a/(a+b)$ and substituting in equation (7.7) yield $p_1 = u^n + (1-u)^n$. Let us conjecture that the inequality $p_1 \geq p_2$ holds for $n > 2$. This conjecture is equivalent to the conjectured inequality

$$u^n + (1-u)^n \geq 1/2 \tag{7.8}$$

The maximum of the expression $x^n + y^n$ where $x + y = 1$ is obtained when $x = y = 1/2$. Consequently, the maximum of the left-hand side of inequality (7.8) is obtained when $u = 1 - u$ or when $u = 1/2$. However, $(1/2)^n + (1/2)^n < 1/2$ for any $n > 2$. The only value for which the equality in inequality (7.8) is attained is $n = 2$. As a result, inequality (7.8) does not hold for $n > 2$.

Now, let us conjecture that the inequality $p_1 \leq p_2$ holds for any $n > 2$. This is equivalent to the inequality:

$$u^n + (1-u)^n \leq 1/2 \tag{7.9}$$

The conjectured inequality (7.9) can be disproved by providing a counterexample. For example, the inequality is not correct for $n = 3$ and $u = 0.9$. By taking u sufficiently close to unity, inequality (7.9) can be disproved for any other $n > 2$.

Furthermore, no inequality similar to inequality (7.6) holds for more than two varieties. Indeed, let a, b and c be the number of items of variety A, B and C in a large batch of items.

The quantity

$$p = a^n / (a+b+c)^n + b^n / (a+b+c)^n + c^n / (a+b+c)^n \tag{7.10}$$

expresses the probability p of selecting n components of the same variety from a large batch ($n \ll a+b+c$) containing three varieties.

Denoting $u = a/(a+b+c)$, $v = b/(a+b+c)$, $w = c/(a+b+c)$ and substituting in equation (7.10) yield $p = u^n + v^n + w^n$. Let us conjecture that the inequality $p \geq 1/2$ holds for $n \geq 2$. This conjecture is equivalent to the conjectured inequality

$$u^n + v^n + w^n \geq 1/2 \tag{7.11}$$

Since $u+v+w=1$, the maximum of the left-hand side of inequality (7.11) is obtained when $u = v = w = 1/3$. However, $(1/3)^n + (1/3)^n + (1/3)^n < 1/2$ for any $n \geq 2$. As a result, for more than two varieties, inequality (7.11) does not hold for any $n \geq 2$.

7.4 UPPER BOUND OF THE PROBABILITY OF SELECTING EACH COMPONENT FROM DIFFERENT VARIETY

Consider the algebraic inequality

$$n!\, x_1 x_2 \ldots x_n \leq n!/n^n \tag{7.12}$$

where $0 \leq x_i \leq 1, \sum_{i=1}^{n} x_i = 1$

This inequality can be proved by using the arithmetic mean-geometric mean inequality, according to which

$$(x_1 x_2 \ldots x_n)^{1/n} \leq \frac{x_1 + x_2 + \cdots + x_n}{n} \tag{7.13}$$

stating that the geometric mean of n positive numbers $x_1, x_2, \ldots, x_n$ is always smaller than or equal to their arithmetic mean. Substituting $\sum_{i=1}^{n} x_i = 1$ in inequality (7.13), raising the positive quantities into the power of 'n' and multiplying both sides by $n!$ give inequality (7.12).

Inequality (7.12) has a useful interpretation. Suppose that in a large batch, there are n different varieties (A_1, A_2,..., A_n) of components where the variety fractions are x_1, x_2,..., x_n $\left(\sum_{i=1}^{n} x_i = 1\right)$, respectively. The left-hand side of inequality (7.12) then gives the upper bound of the probability of selecting n items from different variety because n different varieties can be selected in $n!$ distinct ways.

Consider the case corresponding to $n = 2$ in inequality (7.12):

$$2x_1x_2 \le 2!/2^2 = 1/2 \tag{7.14}$$

The variables x_1 and x_2 are interpreted as fractions of items from two varieties A and B in a large batch of items. Let a and b be the number of items of variety A and variety B in a large batch of items. Let $x_1 = a/(a+b)$ denote the probability of random selection of an item of variety A and $x_2 = b/(a+b)$ — the probability of random selection of an item of variety B.

Suppose that a working assembly can always be made if at least two of the selected components are of the same variety. In this case, it is important to know the maximum possible probability of selecting items from different variety because making such a selection will result in a faulty assembly. From inequality (7.14), it follows that the probability of a faulty assembly from purchasing two components of different variety can never exceed 0.5, irrespective of the proportions of the separate varieties in the batch, which confirms the conclusions from Section 7.3.

This reasoning can be extended for three distinct varieties in the batch. Let a, b and c be the number of items of variety A, B and C, correspondingly, in a large batch of items. Let $x_1 = a/(a+b+c)$ denote the probability of selecting an item of variety A, $x_2 = b/(a+b+c)$ — the probability of selecting an item of variety B and $x_3 = c/(a+b+c)$ — the probability of selecting an item of variety C. Suppose that to make a working assembly, at least two items of the same variety must be selected. It is important to know the maximum possible probability of selecting three items, each of which is of different variety, because only such a selection will result in a faulty assembly.

For $n = 3$, inequality (7.12) results in

$$3!\, x_1x_2x_3 \le 3!/3^3 = 2/9 \tag{7.15}$$

which predicts that the probability of a faulty assembly resulting from selecting all components from different variety never exceeds 2/9 irrespective of the proportions of the separate varieties in the batch.

For n components, for the probability of a faulty assembly resulting from selecting all components from different variety, inequality (7.12) gives an upper bound of $n!/n^n$, irrespective of the proportions of the separate varieties in the batch.

7.5 LOWER BOUND OF THE PROBABILITY OF RELIABLE ASSEMBLY

Consider the algebraic inequality

$$x_1^n + x_2^n + \cdots + x_n^n \ge \frac{1}{n^{n-1}} \tag{7.16}$$

where $0 \le x_i \le 1, \sum_{i=1}^{n} x_i = 1$

Inequality (7.16) can be proved by reducing its complexity through the Chebyshev's sum inequality.

Without loss of generality, it can be assumed that $x_1 \ge x_2 \ge \ldots \ge x_n$. From the basic properties of inequalities it follows that $x_1^{n-1} \ge x_2^{n-1} \ge \ldots \ge x_n^{n-1}$. According to the Chebyshev's sum inequality introduced in Chapter 2 (Section 2.1.6):

$$\frac{x_1^n + x_2^n + \cdots + x_n^n}{n} \ge \frac{x_1^{n-1} + x_2^{n-1} + \cdots + x_n^{n-1}}{n} \times \frac{x_1 + x_2 + \cdots + x_n}{n}$$

$$= \frac{1}{n^2} \times \left(x_1^{n-1} + x_2^{n-1} + \cdots + x_n^{n-1}\right) \tag{7.17}$$

As a result,

$$x_1^n + x_2^n + \cdots + x_n^n \ge \frac{1}{n^1} \times \left(x_1^{n-1} + x_2^{n-1} + \cdots + x_n^{n-1}\right) \tag{7.18}$$

In exactly the same way, the complexity of $x_1^{n-1} + x_2^{n-1} + \cdots + x_n^{n-1}$ in the right-hand side of inequality (7.18) can be reduced:

$$\frac{x_1^{n-1} + x_2^{n-1} + \cdots + x_n^{n-1}}{n} \ge \frac{x_1^{n-2} + x_2^{n-2} + \cdots + x_n^{n-2}}{n} \times \frac{x_1 + x_2 + \cdots + x_n}{n}$$

$$= \frac{1}{n^2} \times \left(x_1^{n-2} + x_2^{n-2} + \cdots + x_n^{n-2}\right)$$

As a result,

$$x_1^n + x_2^n + \cdots + x_n^n \ge \frac{1}{n^2} \times \left(x_1^{n-2} + x_2^{n-2} + \cdots + x_n^{n-2}\right) \tag{7.19}$$

The complexity of $x_1^{n-2} + x_2^{n-2} + \cdots + x_n^{n-2}$ in the right-hand side of inequality (7.19) can, in turn, be reduced by applying the Chebyshev's sum inequality and continuing this process finally leads to

$$x_1^n + x_2^n + \cdots + x_n^n \ge \frac{1}{n^{n-1}} \times \left(x_1^1 + x_2^1 + \cdots + x_n^1\right) = \frac{1}{n^{n-1}} \tag{7.20}$$

which completes the proof of inequality (7.16).

Let $x_1, x_2, \ldots, x_n$ be the fractions of components from varieties $A_1, A_2, \ldots, A_n$, respectively, in a large batch of components.

The left-hand side of inequality (7.16) can be interpreted as the probability of selecting n components of the same variety and the right-hand side of the inequality is the lower bound of this probability.

The obtained lower bound of the probability of selecting components of the same variety can be used to determine the upper bound of the probability of not selecting components of the same variety.

Consider the complementary events A (all selected components are of the same variety) and $\bar{A}$ (not all selected components are of the same variety) whose union is the certain event Ω: $A \cup \bar{A} = \Omega$. Consequently, $P(A) + P(\bar{A}) = 1$. Assessing the upper bound U of the probability of event $\bar{A}$, $P(\bar{A}) \le U$ can be made by using the lower bound L of the probability of the complementary event A:

$$P(\bar{A}) \le U = P(A) \ge L \tag{7.21}$$

Since for complementary events A and $\bar{A}$, the relationship $P(\bar{A}) = 1 - P(A)$ holds, it follows that the upper bound U of $P(\bar{A})$ is equal to 1–' the lower bound L of $P(A)$':

$$U = 1 - L \tag{7.22}$$

As a result, the upper bound U of the probability $P(\bar{A})$ that not all components are of the same variety is given by

$$U = 1 - \frac{1}{n^{n-1}} \tag{7.23}$$

Applying the principle of inversion in calculating the upper bound of the probability $P(\bar{A})$ through the lower bound L of the probability $P(A)$ leads to a significant simplification of models and calculations.

This technique will be demonstrated with a large batch containing three varieties A_1, A_2 and A_3 with fractions x_1, x_2 and x_3. According to inequality (7.20), the lower bound of the probability of selecting all three components of the same variety (event A) is given by

$$P(A) = x_1^3 + x_2^3 + x_3^3 \ge 1/3^2 \tag{7.24}$$

The probability $P(\bar{A})$ that not all components will be of the same variety is given by

$$P(\bar{A}) = 3x_1^2x_2 + 3x_2^2x_1 + 3x_2^2x_3 + 3x_3^2x_2 + 3x_3^2x_1 + 3x_1^2x_3 + 6x_1x_2x_3 \tag{7.25}$$

which is a complex expression. This is a sum of the probabilities of selecting two components of variety A_1 and one component of another variety, selecting two components of variety A_2 and one component of another variety, selecting two components of variety A_3 and one component of another variety and selecting three components of three different varieties. As can be verified, $P(A) + P(\bar{A}) = 1$ because

$$x_1^3 + x_2^3 + x_3^3 + 3x_1^2x_2 + 3x_2^2x_1 + 3x_2^2x_3 + 3x_3^2x_2 + 3x_3^2x_1 + 3x_1^2x_3$$
$$+ 6x_1x_2x_3 = (x_1 + x_2 + x_3)^3 = 1 \tag{7.26}$$

According to equation 7.23, the upper bound U of the probability (7.26) is given by

$$P(\bar{A}) \leq 1 - 1/3^2 = 8/9 \qquad (7.27)$$

These results have been confirmed by Monte Carlo simulation.

7.6 TIGHT LOWER AND UPPER BOUND FOR THE FRACTION OF FAULTY COMPONENTS IN A POOLED BATCH

Consider the set of ordered positive fractions $0 \leq p_1 \leq p_2 \leq \ldots \leq p_m \leq 1$. If the fractions p_i are presented as a ratio $p_i = a_i / n_i$ of two integer numbers a_i and n_i, ($i = 1, \ldots, n$), the following algebraic inequalities hold:

$$\frac{a_1}{n_1} \leq \frac{a_1 + a_2 + \cdots + a_m}{n_1 + n_2 + \cdots + n_m} \leq \frac{a_m}{n_m} \qquad (7.28)$$

where $a_1 / n_1 \leq a_2 / n_2 \leq \ldots \leq a_m / n_m$.

Inequalities (7.28) can be proved by a mathematical induction. For the trivial case $n = 2$, it can be shown that if $0 \leq p_1 \leq p_2 \leq 1$, $p_1 = a_1 / n_1$ and $p_2 = a_2 / n_2$, then

$$\frac{a_1}{n_1} \leq \frac{a_1 + a_2}{n_1 + n_2} \leq \frac{a_2}{n_2} \qquad (7.29)$$

Indeed, from $p_1 \leq p_2$, it follows that $a_1 / n_1 \leq a_2 / n_2$ which is equivalent to

$$a_1 n_2 \leq a_2 n_1 \qquad (7.30)$$

Adding the quantity $a_1 n_1$ to both sides of inequality (7.30) results in $a_1 n_2 + a_1 n_1 \leq a_2 n_1 + a_1 n_1$ which, after factoring a_1 from the left-hand side and n_1 from the right-hand side, results in

$$a_1 (n_1 + n_2) \leq n_1 (a_1 + a_2) \qquad (7.31)$$

Dividing both sides of inequality (7.31) by the positive value $n_1 (n_1 + n_2)$ does not alter the direction of the inequality. The result is the inequality

$$p_1 = \frac{a_1}{n_1} \leq \frac{a_1 + a_2}{n_1 + n_2} \qquad (7.32)$$

which is the left inequality from inequalities (7.28), for $m = 2$.

The right inequality from inequalities (7.28), for $m = 2$, can be proved in a similar fashion.

Suppose now that inequalities (7.28) hold for the integer k, where $k \geq 2$, $0 \leq p_1 \leq p_2 \leq \ldots \leq p_k \leq 1$, $p_i = x_i / n_i$ (induction hypothesis):

$$\frac{a_1}{n_1} \leq \frac{a_1 + a_2 + \cdots + a_k}{n_1 + n_2 + \cdots + n_k} \leq \frac{a_k}{n_k} \tag{7.33}$$

Consider again the left inequality

$$\frac{a_1}{n_1} \leq \frac{a_1 + a_2 + \cdots + a_k}{n_1 + n_2 + \cdots + n_k} \tag{7.34}$$

Multiplying both sides of inequality (7.34) by the positive quantity $n_1(n_1 + n_2 + \cdots + n_k)$ does not alter its direction and the equivalent inequality

$$a_1(n_1 + n_2 + \cdots + n_k) \leq n_1(a_1 + a_2 + \cdots + a_k) \tag{7.35}$$

is obtained. Without loss of generality, suppose that for the k + 1st term $p_{k+1} = \frac{a_{k+1}}{n_{k+1}}$, the ranking

$$0 \leq p_1 \leq p_2 \leq \ldots \leq p_k \leq p_{k+1} \leq 1 \tag{7.36}$$

holds. (If this is not the case, the terms p_i can always be renumbered so that inequality (7.36) is fulfilled.) Since $p_1 = \frac{a_1}{n_1} \leq p_{k+1} = \frac{a_{k+1}}{n_{k+1}}$, it follows that

$$a_1 n_{k+1} \leq n_1 a_{k+1} \tag{7.37}$$

If inequality (7.35) is added to inequality (7.37), the inequality

$$a_1(n_1 + n_2 + \cdots + n_k + n_{k+1}) \leq n_1(a_1 + a_2 + \cdots + a_k + a_{k+1}) \tag{7.38}$$

is obtained which is equivalent to

$$a_1 / n_1 \leq (a_1 + a_2 + \cdots + a_k + a_{k+1}) / (n_1 + n_2 + \cdots + n_k + n_{k+1}) \tag{7.39}$$

With the trivial case, corresponding to $k = 2$ and the proved induction step (7.39), according to the principle of the mathematical induction, the left inequality (7.28) holds for any $m > k$.

In a similar fashion, the right inequality (7.28) can be proved.

Inequality (7.28) has a useful interpretation. Suppose that a_i $(i = 1,\ldots,n)$ stands for the number of faulty components in the ith batch and n_i $(i = 1,\ldots,n)$ stands for the total number of components in the ith batch and $p_i = a_i / n_i$ $(i = 1,\ldots,n)$ stands for the fraction of faulty items in the ith batch. The number of components n_i $(i = 1,\ldots,n)$ in the separate batches *is unknown*.

The batch with the smallest fraction of faulty components will be referred to as the 'best batch' and the batch with the largest fraction of faulty components will be referred to as the 'worst batch'. Now, a meaningful interpretation of inequality (7.28) can be provided. The inequality predicts that if n batches containing known fractions of defective components are pooled into a single batch, irrespective of the number of components n_i in the separate batches, the percentage of faulty components in the pooled batch is always smaller than the percentage of faulty components in the worst batch and larger than the percentage of faulty components in the best batch.

The fraction of faulty items in the pooled batch always remains within the tight bounds p_1 and p_m:

$$p_1 \leq p \leq p_m \tag{7.40}$$

where $p = \dfrac{a_1 + a_2 + \cdots + a_m}{n_1 + n_2 + \cdots + n_m}$.

This example demonstrates determining tight lower and upper bound for the percentage of faulty items in the pooled batch, without any knowledge related to the sizes of the constituent batches.

7.7 AVOIDING AN OVERESTIMATION OF EXPECTED PROFIT

7.7.1 Avoiding the Risk of Overestimating Profit through Interpretation of the Jensen's Inequality

Consider the concave function $f(x)$ for which the following Jensen's inequality holds

$$f(w_1x_1 + w_2x_2 + \cdots + w_nx_n) \geq w_1f(x_1) + w_2f(x_2) + \cdots + w_nf(x_n) \tag{7.41}$$

where w_i $(i = 1,\ldots,n)$ are weights that satisfy $0 \leq w_i \leq 1$ and $w_1 + w_2 + \cdots + w_n = 1$.

If the weights are chosen to be equal, $w_i = 1/n$, the Jensen's inequality (7.41) becomes

$$f\left((1/n)\sum_{i=1}^{n} x_i\right) \geq (1/n)\sum_{i=1}^{n} f(x_i) \tag{7.42}$$

The demand for a particular product X is almost always associated with variation (X is a random variable) and the variation of X cannot be controlled. The profits Y depend on the demand X through a particular function $Y = f(X)$.

Now, suppose that the variables x_i in inequality (7.42) are interpreted as 'levels of demand' for the product and $f(x_i)$ are the profits corresponding to these levels of demand.

Inequality (7.42) predicts that the average of the profits at different levels of the demand is smaller than the profit calculated at an average level of the demand. The inequality helps the decision-maker choose between two competing strategies:

a. Averaging first different values of the demand X, $\overline{x} = (1/n)\sum_{i=1}^{n} x_i$, by using n random values $x_1, x_2, \ldots, x_n$ of the demand within the demand range, followed by assessing the average profit from $\overline{y} = f(\overline{x})$.
b. Obtaining the average profit $\overline{y} = (1/n)\sum_{i=1}^{n} f(x_i)$ by averaging the profits corresponding to n random demands $x_1, x_2, \ldots, x_n$ within the demand range.

Inequality (7.42) effectively states that strategy '*a*' overestimates the profit.

To demonstrate the significant difference between the two decision strategies, consider an example from the biotech industry, where the demand x for a particular biochemical product varies from 0 to 300e3 kg/year and the capacity of the production plant for 1 year is only 200e3 kg of product per year. Suppose that the profit generated from selling the product if the demand is in the interval (0,200e3) is given by $f(x) = 3.6x$, where x is the quantity of the product [in kg] sold. The profit generated from selling the product if the demand is in the interval (200e3,300e3) is given by $f(x) = 3.6 \times 200\text{e}3$ because the production capacity cannot exceed 200e3 kg.

The profit function is, therefore, a concave function (Figure 7.2) defined in the following way:

$$f(x) = \begin{cases} 3.6x, & 0 \le x \le 200\text{e}3 \\ 3.6 \times 200\text{e}3, & 200e3 \le x \le 300\text{e}3 \end{cases} \tag{7.43}$$

The average demand is obviously 300e3/2 = 150e3 kg/year. The profit corresponding to the average demand is $\overline{y}_1 = 3.6 \times 150\text{e}3 = 540\text{e}3$. This is the value $\overline{y}_1 = f\left((1/n)\sum_{i=1}^{n} x_i\right)$ on the left-hand side of inequality (7.42).

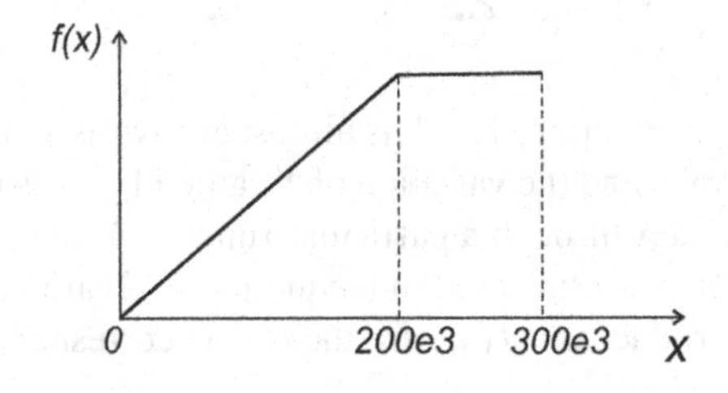

FIGURE 7.2 Concave profit-demand dependence generated from selling a product.

The average of the profits was calculated by using a simple Monte Carlo simulation.

Running the Monte Carlo simulation with 100,000 trials resulted in an average profit equal to $\overline{y}_2 = 480e3$. This is the value $\overline{y}_2 = (1/n)\sum_{i=1}^{n} f(x_i)$ on the right-hand side of inequality (7.42). The difference between the two values for the average profit is significant. Because of the significant difference, critical business decisions should not be made on the basis of profit taken at the average demand. The correct decision strategy that eliminates the risk of overestimating the average profit is to average the profits at different levels of the demand rather than taking the profit at the average level of demand.

7.7.2 Avoiding Overestimation of the Average Profit through Interpretation of the Chebyshev's Sum Inequality

Avoiding the overestimation of the average profit can also be illustrated with the interpretation of the Chebyshev's sum inequality, proved in Chapter 2.

Let $a_1,\ldots,a_n$ and $b_1,\ldots,b_n$ be two sets of similarly ordered positive numbers $(a_1 \leq,\ldots,\leq a_n)$ and $(b_1 \leq,\ldots,\leq b_n)$. Then, the Chebyshev's sum inequality holds:

$$\frac{a_1 b_1 + \cdots + a_n b_n}{n} \geq \frac{a_1 + \cdots + a_n}{n} \cdot \frac{b_1 + \cdots + b_n}{n} \tag{7.44}$$

If the number sequences are oppositely ordered, for example, if $a_1 \geq,\ldots,\geq a_n$ and $b_1 \leq,\ldots,\leq b_n$, the inequality is reversed.

$$\frac{a_1 b_1 + \cdots + a_n b_n}{n} \leq \frac{a_1 + \cdots + a_n}{n} \cdot \frac{b_1 + \cdots + b_n}{n} \tag{7.45}$$

Suppose that in inequality (7.45), the variables a_i $(i = 1,\ldots,n)$, ranked in descending order $(a_1 \geq,\ldots,\geq a_n)$, stand for probabilities of a gain (return) from risky investments. The variables b_i $(i = 1,\ldots,n)$, ranked in ascending order $(b_1 \leq,\ldots,\leq b_n)$, stand for the returns from the risky investments which correspond to the probabilities a_i. This negative correlation between the probabilities and returns from risky investments *is very common*: the smaller the likelihood of success characterising a risky investment, the larger the return from the investment is.

The right-hand side of inequality (7.45) can now be interpreted as an estimate of the expected profit per investment by using the average return from the investments $\overline{b} = (b_1 + b_2 + \cdots + b_n)/n$ and the average probability of return $\overline{a} = (a_1 + \cdots + a_n)/n$. The left-hand side of inequality (7.45) can be interpreted as the expected profit per investment, assessed by taking the expected profit from each individual investment.

Inequality (7.45) predicts that estimating the expected return from an investment by using the average return and the average probability of return leads to an overestimation of the expected return from an investment.

The overestimation of the expected potential return from an investment can be significant, and this will be illustrated by a very simple numerical example involving only two investments: an investment with potential return of $c_1 = \$800$ and an investment with potential return of $c_2 = \$15{,}000$. The probabilities of return from the investments are $p_1 = 0.77$ and $p_2 = 0.60$, correspondingly. If the average probability of return from an investment and the average return are used for calculating the expected potential return per investment, the value $\bar{p} \times \bar{c} = 0.5(0.77 + 0.60) \times 0.5(800 + 15{,}000) = 5{,}411$ for the expected potential return will be predicted. The actual expected potential return per investment is

$$\frac{p_1c_1 + p_2c_2}{2} = \frac{0.77 \times 800 + 0.60 \times 15{,}000}{2} = 4{,}808$$

which is significantly smaller than the prediction of 5,411, based on the average values of the probabilities and the returns.

The temptation to work with average values of the probability of return and the returns is greater if the risky portfolio contains a large number of investments. Such is, for example, the case with an agency making profit from lending to business customers. The profit margin from charging interest rate on loans to customers is negatively correlated with the size of the customer's loan. Business customers with a small loan are small or have poorer credit rating. For this reason, the charged interest is higher and the profit margin is higher. Business customers with large loans are large or have a better credit rating, hence, the charged interest is lower and the profit margin is lower.

In the Chebyshev's sum inequality (7.45), $a_1 \geq, \ldots, \geq a_n$ now stand for the specified profit margins corresponding to each of n customer loans represented by the variables b_i $(i = 1, \ldots, n)$, ranked in ascending order: $b_1 \leq, \ldots, \leq b_n$.

The right-hand side of inequality (7.45) can be interpreted as the projected expected profit $\overline{ab}$ per customer, made by using the average size of the loans $\bar{b} = (b_1 + b_2 + \cdots + b_n) / n$ and the average profit margin $\bar{a} = (a_1 + \cdots + a_n) / n$. The left-hand side of inequality (7.45) can be interpreted as the actual average profit per customer. The Chebyshev's inequality (7.45) now predicts that the projected expected profit per customer, estimated by using the average loan $\bar{b} = (b_1 + b_2 + \cdots + b_n) / n$ and the average profit margin $\bar{a} = (1 / n)(a_1 + \cdots + a_n)$, is overestimated.

The overestimation of the projected expected profit can be significant, and this will be illustrated by a simple numerical example involving two customers only, with loans $b_1 = \$600$ and $b_2 = \$40{,}000$. The profit margins are $a_1 = 0.12$ and $a_2 = 0.04$, correspondingly. If the average profit margin and the average loan are used for calculating the projected average profit per customer, the value $\bar{a} \times \bar{b} = 0.5(0.12 + 0.04) \times 0.5(600 + 40{,}000) = 1{,}624$ will be predicted. The actual average profit per customer is $\dfrac{p_1b_1 + p_2b_2}{2} = \dfrac{0.12 \times 600 + 0.04 \times 40{,}000}{2} = \$836$ which is significantly smaller than the predicted value.

The conclusion is that for negatively correlated variables, calculating the average by multiplying the averages of the two variables leads to an overestimation.

Conversely, according to inequality (7.44), for positively correlated variables, calculating the average by multiplying the averages of the two variables leads to an underestimation.

7.7.3 Avoiding Overestimation of the Probability of Successful Accomplishment of Multiple Tasks

Consider the inequality proved in Section 2.1.3 of Chapter 2:

$$\left(\frac{m_1x_1 + m_2x_2 + \cdots + m_nx_n}{M}\right)^M \geq x_1^{m_1}x_2^{m_2}\ldots x_n^{m_n} \tag{7.46}$$

where $m_1,\ldots,m_n$ and $x_1,\ldots,x_n$ are positive values, and $M = \sum_{i=1}^{n} m_i$.

Let x_i $(0 < x_i < 1)$ stand for the probability of successful accomplishment of a task of type i, where $i = 1,2,\ldots,n$. The accomplishment of any tasks is an event statistically independent from the accomplishment of any other task.

Let m_i stand for the number of tasks of type i, where $i = 1,2,\ldots,n$.

The right-hand side of inequality (7.46) stands for the probability of successful accomplishment of all $M = \sum_{i=1}^{n} m_i$ tasks.

Indeed, the probability of successful accomplishment of m_1 tasks of type one is $x_1^{m_1}$, the probability of successful accomplishment of m_2 tasks of type two is $x_2^{m_2}$ and so on. The probability of successful accomplishment of all n types of tasks is, therefore, given by the right-hand side of inequality (7.46).

The expression $\overline{x} = \frac{m_1x_1 + m_2x_2 + \cdots + m_nx_n}{M}$ on the left-hand side of inequality (7.46) stands for the average probability $\overline{x}$ of successful accomplishment of a task, irrespective of its type. It is simply obtained by adding the probabilities characterising the separate tasks and dividing it by the total number $M = \sum_{i=1}^{n} m_i$ of tasks. Inequality (7.46) can be rewritten as

$$\overline{x}^M \geq x_1^{m_1}x_2^{m_2}\ldots x_n^{m_n} \tag{7.47}$$

Inequality (7.47) predicts that the probability of successful accomplishment of all M tasks by using the average probability of successful accomplishment of a task is greater than the actual probability of successful accomplishment of all M tasks. Therefore, using the average probability of successful accomplishment of a task leads to an overestimation of the actual probability of successful accomplishment of all tasks.

8 Generating New Knowledge by Interpreting Algebraic Inequalities in Terms of Potential Energy

8.1 INTERPRETING AN INEQUALITY IN TERMS OF POTENTIAL ENERGY

Consider the general algebraic inequality

$$f(x_1, x_2, \ldots, x_n) \geq L \tag{8.1}$$

where $x_1, x_2, \ldots, x_n$ are system/process parameters and L is an unknown lower bound. The parameters may be subjected to a constraint:

$$\varphi(x_1, x_2, \ldots, x_n) = 0$$

where $\varphi(x_1, x_2, \ldots, x_n) = 0$ is an arbitrary continuous function of the system/process parameters $x_1, x_2, \ldots, x_n$. A common constraint is $\varphi \equiv \sum_{i=1}^{n} x_i = d$, where d is a constant.

Often, the function $f(x_1, x_2, \ldots, x_n)$ can be interpreted as *potential energy* of the system. If this can be done, the constant L in the right-hand side of inequality (8.1) can be interpreted as the minimum potential energy of the system which corresponds to its state of stable equilibrium. From the equilibrium conditions that correspond to the stable equilibrium, a number of useful relationships can be derived and the lower bound L determined without resorting to complex models.

To illustrate the idea behind the potential energy approach (Todinov, 2020e), a simple example will be used. Consider the inequality:

$$\sqrt{a^2 + x^2} + \sqrt{b^2 + y^2} \geq L \tag{8.2}$$

where the non-negative parameters x, y are subjected to the constraint

$$x + y = d$$

DOI: 10.1201/9781003199830-8

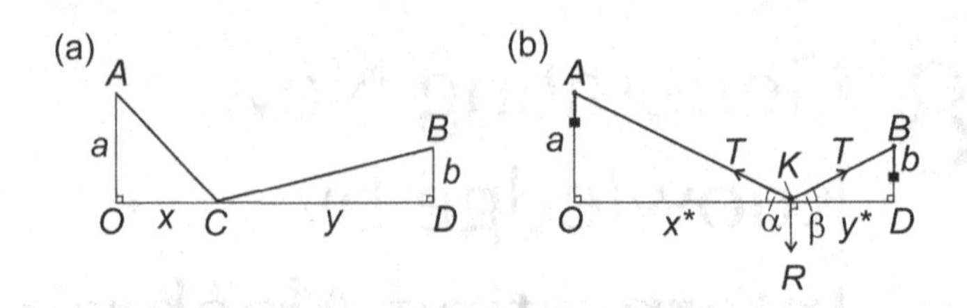

FIGURE 8.1 Meaningful interpretation of inequality (8.1) in terms of potential energy of a constant-tension spring: (a) potential energy corresponding to an arbitrary location of point C; (b) location of the ring K which corresponds to a minimal potential energy of the system.

where $d = OD$ (Figure 8.1a) and a, b are given constants. In inequality (8.2), the lower bound L is unknown quantity.

Noticing that both $\sqrt{a^2 + x^2}$ and $\sqrt{a^2 + y^2}$ can be interpreted as a hypotenuse of a right-angle triangle, inequality (8.2) can be presented by using the two right-angle triangles OAC and CBD in Figure 8.1a.

The left-hand side of inequality (8.2) is the length of ACB and the lower bound L is the smallest value of the length ACB.

In interpreting the left-hand side of inequalities (8.1) and (8.2) as potential energy, physical analogies such as *constant-tension springs*, *zero-length linear springs* and *zero-length non-linear springs* prove to be useful.

A *constant-tension* spring is a spring whose tension is independent of its length (Levi, 2009). It is assumed that the tension T in the constant-tension spring, when the system is in equilibrium, is equal to unity ($T = 1$).

The potential energy E of a constant-tension spring of length x is given by

$$E = \int_0^x T\,du = Tx = x \tag{8.3}$$

which is product of the spring length x and the tension of the spring T (T=1).

Note that the potential energy of a constant-tension spring ($E = x$) as a function of the spring length x is different from the potential energy $E = (1/2)kx^2$ of a conventional spring stretched to a length x. This is because, for a conventional linear spring, the spring force $F = kx$ varies proportionally with the displacement x, while for a constant-tension spring, the spring force is constant $F = 1$ and does not depend on the displacement x. A constant-tension spring can be constructed with weights of unit magnitude and pulleys (Figure 8.1b). For constant weights, the spring force is always constant. The potential energy, however, varies with the spring length. The larger the length $KA + KB$ of the constant-tension spring (Figure 8.1b), the larger is the sum of the elevations of the suspended unit weights and the larger is the potential energy of the system.

Suppose that K in Figure 8.1b is a ring that moves without any friction along the segment OD, whose length is equal to d and AKB is a non-stretchable string kept under constant tension through the pulleys A and B and the unit weights at the ends. The left part of inequality (8.2) is then proportional to the potential energy of the system in Figure 8.1b. The smallest potential energy of the system in Figure 8.1b corresponds to the smallest length $KA + KB$ of the string, which marks the smallest combined elevation of the unit weights.

The constant L on the right part of inequality (8.3) represents the minimum potential energy of the system. For a state of stable equilibrium, the forces applied to point K on the string must balance out. Because the ring is frictionless, the force R applied to the string from the ring is perpendicular to the x-axis and its component along the x-axis is zero. Consequently, the sum of the components of the tension force T applied on the string at point K, along the x-axis, must be equal to zero ($-T\cos\alpha + T\cos\beta = 0$). From this necessary condition for a system equilibrium, it is clear that the angle AKO (α) must be equal to the angle BKD (β). In this case, triangles AKO and BKD are similar and $x^*/a = (d - x^*)/b$. From this relationship, $x^* = ad/(a+b)$ and $y^* = bd/(a+b)$. The value of the lower bound L in inequality (8.2) is equal to the smallest length $L = (a+b)\sqrt{1+d^2/(a+b)^2}$ of AKB. As a result, the exact value of the lower bound L in inequality (8.2) has been determined solely on the basis of a meaningful interpretation of inequality (8.2) in terms of potential energy.

8.2 A NECESSARY CONDITION FOR MINIMISING SUM OF THE POWERS OF DISTANCES

Consider n points in space: $A_1, A_2,\ldots, A_n$ and an extra point M with distances $r_1, r_2,\ldots, r_n$ to the points $A_1, A_2,\ldots, A_n$, respectively (Figure 8.2a).

Consider the algebraic inequality (8.4) involving the distances $r_1, r_2,\ldots, r_n$:

$$r_1^n + r_2^n + \cdots + r_n^n \geq L \tag{8.4}$$

where L is the lower bound which is unknown quantity.

Suppose also that the connecting segments $r_i = M_0A_i$ (Figure 8.2b) are *non-linear zero-length tension springs* of order $n-1$. A non-linear zero-length spring of order $n-1$ is a spring whose tension T is directly proportional to the $n-1$ power of its length r ($T = kr^{n-1}$). The potential energy u of a stretched to a length r non-linear zero-length spring of order n-1, is given by

$$u = \int_0^r kv^{n-1}dv = kr^n / n \tag{8.5}$$

From equation 8.5, it can be seen that if the constant k is set to be equal to n, the left part of inequality (8.4) can be interpreted as the total potential energy of a system of n non-linear zero-length springs of order $n-1$, while the right part of

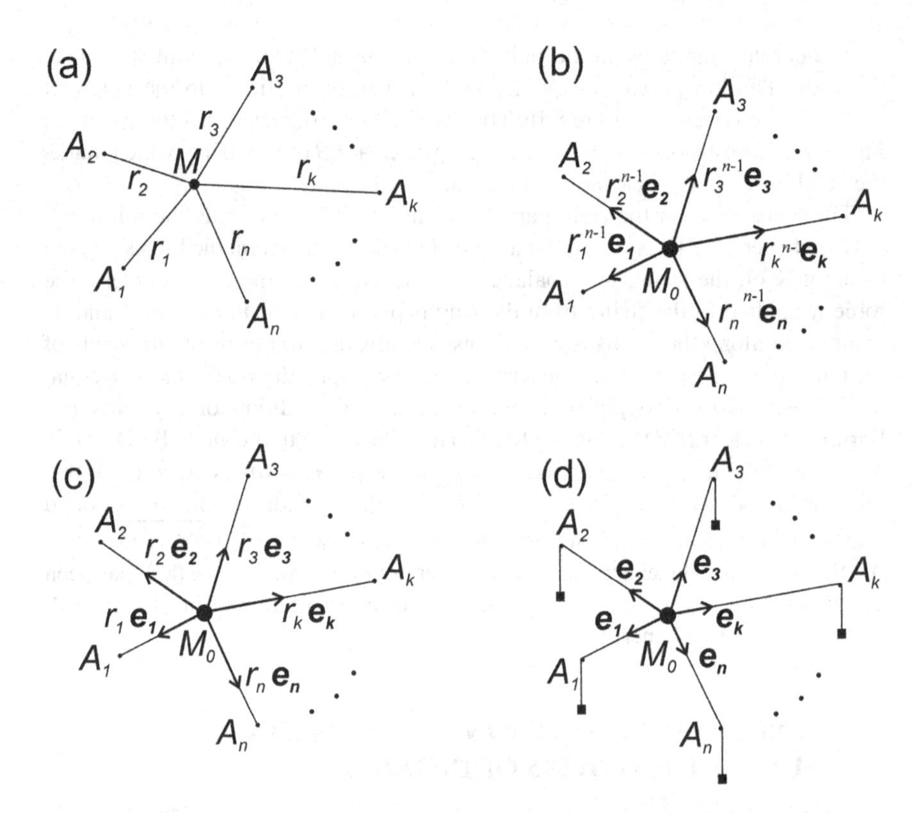

FIGURE 8.2 (a) Distances MA_i from point M to a specified set of points in space; (b) the necessary condition for a minimal sum of distances M_0A_i raised to the power of n is vectors $r_i^{n-1}e_i$ to add up to zero; (c) the necessary condition for a minimal sum of squared distances M_0A_i is point M_0 to be located at the mass centre of the system of points; (d) the necessary condition for a minimal sum of distances M_0A_i is point M_0 to be located at the geometric median of the system of points.

inequality (8.4) can be interpreted as the minimum possible potential energy of the system of non-linear zero-length springs of order $n-1$.

If point M is now selected at such a location (point M_0 in Figure 8.2b) that the sum of the nth power of the distances $r_1, r_2, \ldots, r_n$ is a minimum, the equality

$$r_1^n + r_2^n + \cdots + r_n^n = L \tag{8.6}$$

will hold.

Since the minimum potential energy corresponds to a stable equilibrium of the system, the sum of the vectors $r_i^{n-1}\mathbf{e}_i$ with origin point M_0, directed towards the separate points $A_1, A_2, \ldots, A_n$, must be zero (Figure 8.2b):

$$\sum_{i=1}^{n} r_i^{n-1}\mathbf{e}_i = 0 \tag{8.7}$$

As can be seen from Figure 8.2b, the meaningful interpretation of the left and right part of inequality (8.4) led to a necessary condition regarding the optimal location M_0. The point M_0 that corresponds to the minimal sum of distances to the specified points A_i raised into power of n, coincides with the point for which $\sum_{i=1}^{n} r_i^{n-1}\mathbf{e}_i = 0$. There are two important special cases of inequality (8.4) which correspond to the cases $n = 2$ and $n = 1$.

8.3 DETERMINING THE LOWER BOUND OF THE SUM OF SQUARED DISTANCES TO A SPECIFIED NUMBER OF POINTS IN SPACE

Consider the special case of inequality (8.4) where $n = 2$. In this case, inequality (8.4) transforms into the inequality

$$r_1^2 + r_2^2 + \cdots + r_n^2 \geq L \tag{8.8}$$

The connecting segments $r_i = M_0A_i$ in Figure 8.2c can be interpreted as *zero-length linear tension springs*. The tension T of a linear zero-length tension spring is directly proportional to its length r ($T = kr$). According to equation 8.5, the potential energy of a stretched to a length r linear spring is given by $(1/2)kr^2$. If the spring stiffness k is selected to be equal to 2 ($k = 2$), the left part of inequality (8.8) is the potential energy of a system of n zero-length linear springs while the right part of inequality (8.8) corresponds to the minimum potential energy of the system of zero-length linear tension springs. Since the minimum potential energy corresponds to a stable equilibrium of the system, the sum of the vectors $r_i\mathbf{e}_i$ from point M_0 directed towards the separate points $A_1, A_2, \ldots, A_n$ must be zero: $\sum_{i=1}^{n} r_i\mathbf{e}_i = 0$ (see Figure 8.2c). This means that, in this case, point M_0 must coincide with the mass centre of the system of points.

As can be seen from Figure 8.2c, the interpretation of the left and right part of inequality (8.4) for $n = 2$, led to a necessary condition regarding the location of point M_0. The point with the minimal sum of squared distances to the fixed points coincides with the mass centre of the system of points.

8.4 A NECESSARY CONDITION FOR DETERMINING THE LOWER BOUND OF SUM OF DISTANCES

Consider now the special case of inequality (8.4), where $n = 1$. In this case, inequality (8.4) transforms into the inequality

$$r_1 + r_2 + \cdots + r_n \geq L \tag{8.9}$$

Inequality (8.9) can also be interpreted in a meaningful way. Suppose that the connecting segments M_0A_i are constant-tension springs (Figure 8.2d). Assume that the tension T in each of these springs, when the system is in equilibrium, is equal to unity ($T = 1$).

According to equation 8.3, the potential energy of a stretched to a length r constant-tension spring is given by kr, which is product of the spring length and the tension of the spring ($k = T = 1$). The constant-tension string has been constructed in Figure 8.2d by the weights of unit magnitude and pulleys. The left-hand side of inequality (8.9) is then proportional to the potential energy of the system in Figure 8.2d. Indeed, the larger the sum of distances M_0A_i to the individual points is, the larger is the sum of the elevations of the suspended unit masses and the larger is the potential energy of the system. A minimum potential energy of the system is attained if and only if the sum of the distances M_0A_i to the points is minimal (Figure 8.2d).

Since the minimal potential energy corresponds to a stable equilibrium of the system, the sum of the unit vectors e_i from point M_0 directed towards the separate points, must be zero: $\sum_{i=1}^{n} e_i = 0$ (see Figure 8.2d). As a result, the interpretation of the left- and right-hand side of inequality (8.9) leads to a necessary condition regarding the location M_0. The point with minimal sum of distances to the points $A_1, A_2, \ldots, A_n$ is a point for which the sum of the unit *line-of-sight* vectors towards the points is zero: $\sum_{i=1}^{n} e_i = 0$. In this case, the optimal location M_0 coincides with the *geometric median* (Wesolowsky, 1993) of the system of points. Finding the location M_0 that minimises the sum of the distances to the specified points is the famous *Fermat–Weber location problem* (Chandrasekaran and Tamir, 1990). Algorithms for determining the location M_0 have been presented in Weiszfeld (1937), Chandrasekaran and Tamir (1990) and Papadimitriou and Yannakakis (1982).

The described approach to minimising the sum of distances can be applied not only for a set of points. In Figure 8.3, the circle λ with radius r and coordinates (d, b) of the centre C ($d = OD$, $b = CD$) is inside the angle DOE whose size will be denoted by γ. It is required to select a point K on OD (the x-axis) such that AKB, which is the sum of the distance KA to OE and the distance KB to the circle λ, is

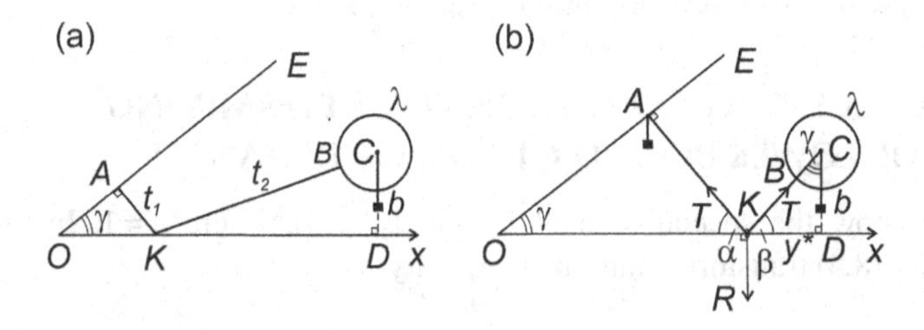

FIGURE 8.3 Minimising the sum of distances of point K to OE and the circle λ: (a) for an arbitrary location of point K; (b) for a location of point K which corresponds to the minimal potential energy of the system.

minimal. It is not difficult to observe that the smallest length AKB corresponds to the smallest length AKC. Therefore, AKB is minimised whenever AKC is minimised. The task consists of determining the lower bound L of

$$AK + KC \geq L \tag{8.10}$$

Suppose again that in Figure 8.3b, K is a ring that moves without any friction along the segment OD whose length is equal to d and AKB is a non-stretchable string kept under constant tension by the pulleys A and C and the unit weights at the ends.

The sum AKC is then proportional to the potential energy of the system in Figure 8.3b. The smallest potential energy of the system in Figure 8.3b corresponds to the smallest length $KA + KC$ of the string which marks the smallest combined elevation of the unit weights. This is also the state of equilibrium of the system. From the necessary condition for equilibrium of the ring K, it is clear that the angle CKD must be equal to the angle AKO. This means that the angle KCD must be equal to γ, $KC = b / \cos\gamma$ and $KB = b / \cos\gamma - r$, the optimal distance DK is obtained from $DK = b\tan(\gamma)$. Since $OK = d - b\tan\gamma$, $AK = (d - b\tan\gamma)\sin\gamma$. The minimal length AKB is obtained from $AKB = b / \cos\gamma - r + (d - b\tan\gamma)\sin\gamma$.

Reframing the question in terms of potential energy significantly simplified the solution.

8.5 A NECESSARY CONDITION FOR DETERMINING THE LOWER BOUND OF THE SUM OF SQUARES OF TWO QUANTITIES

A simple but important illustration of the potential energy interpretation can be given with the inequality

$$a^2 + b^2 \geq L \tag{8.11}$$

where a and b are two non-negative quantities whose sum is equal to a given constant d: $d = a + b$, and L is the lower bound which is unknown quantity.

Suppose that in Figure 8.4, $PD = a$ and $DE = b$ are linear zero-length springs, P is a ring that moves without any friction along the side AC whose length is equal to $d = a + b$, E is a ring that moves without any friction along the side BC whose length is also equal to $d = a + b$ and D is a ring that moves without any friction along AB. The segment AB is a hypotenuse of the isosceles right-angled triangle ABC and the angles EDB and DDA remain constantly equal to $\pi / 4$.

The left-hand side of inequality (8.11) is equal to the total potential energy of the system of linear springs in Figure 8.4a. The smallest potential energy of the system in Figure 8.4b corresponds to the equilibrium position of the rings P, E and D. At the equilibrium position of ring D, the sum of the projections of forces F_1 and F_2 acting on the ring D must be equal to zero ($-F_1\cos(\pi / 4) + F_2\cos(\pi / 4) = 0$) which is only possible if $F_1 = F_2$.

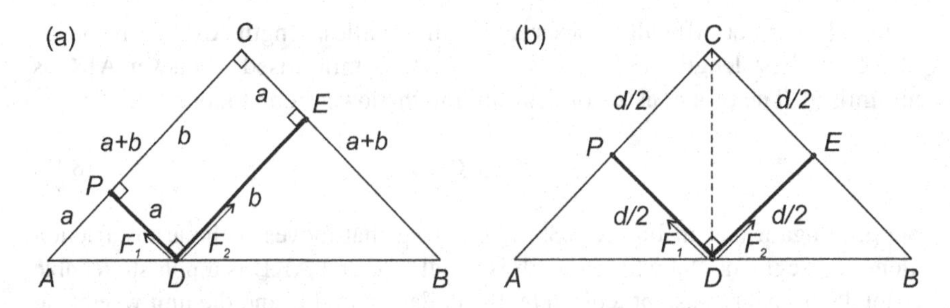

FIGURE 8.4 (a) Squared sum $a^2 + b^2$ which corresponds to an arbitrary location of point D and (b) location of point D which corresponds to the minimal squared sum $a^2 + b^2$

Since the linear spring forces are proportional to the spring lengths ($F_1 = ka$; $F_2 = kb$), this means that if the system is in equilibrium, the ring D must be in the middle of the segment AB. In this position, DP=DE=$d/2$, $|F_1| = |F_2|$, and for a unit spring stiffness $k = 1$, the minimum potential energy $U_{\min}$ of the system becomes

$$U_{\min} = (1/2)(d/2)^2 + (1/2)(d/2)^2 = d^2/4 = (a+b)^2/4$$

Since the potential energy of the system in Figure 8.4a at an arbitrary location of the ring D is $a^2/2 + b^2/2$, the following inequality holds

$$a^2/2 + b^2/2 \geq U_{\min} = d^2/4 \tag{8.12}$$

Multiplying inequality (8.12) by 2 gives

$$a^2 + b^2 \geq d^2/2 \tag{8.13}$$

Consequently, for the constant L in inequality (8.11), $L = d^2/2$ is obtained. The lower bound L in inequality (8.11) is half of the squared sum of the quantities a and b. The exact value of the lower bound L in inequality (8.11) has been determined solely by interpreting inequality (8.11) in terms of potential energy.

8.6 A GENERAL CASE INVOLVING A MONOTONIC CONVEX FUNCTION

The discussion presented in the previous sections related to determining a lower bound by interpreting inequalities as potential energy of springs can be extended in the general case of non-linear springs whose potential energy is defined by a general function $f(x)$. The constraints imposed on the function $f(x)$ are the function to be non-negative and strictly increasing or strictly decreasing in a specified interval $0 \leq x \leq d$, and the function to be convex in the same interval $0 \leq x \leq d$. A convex function means that the function is differentiable in the interval $0 \leq x \leq d$

and its second derivative is greater than zero within that interval ($f''(x) > 0$, $0 \le x \le d$). Consider the inequality

$$f(x) + f(y) \ge L \tag{8.13}$$

where $0 \le x \le d$, $0 \le y \le d$ and $x + y = d$ are constraints and L is unknown lower bound.

By interpretation of inequality (8.13) in terms of potential energy of a system of springs, it can be shown that the lower bound L is given by $L = 2f(d/2)$.

Consider two zero-length non-linear springs AE and BE whose potential energy as a function of their length is given by $U = f(x)$. The total potential energy T of the system of springs in Figure 8.5, as a function of the distance x of point E from the point O (Figure 8.5), is given by the left-hand side of inequality (8.13)

$$T(x) = f(x) + f(d - x) \tag{8.14}$$

where $f(x) \ge 0$ for $x \ge 0$.

Differentiating $T(x)$ with respect to x gives $T'(x) = f'(x) - f'(d - x)$. At $x = d/2$, $T'(x) = 0$; therefore, the point $x = 0$ is a stationary point. The second derivative of the total potential energy is given by $T''(x) = f''(x) + f''(d - x)$. Since $f''(x) > 0$, by the choice of $f(x)$ as a convex function, it follows that $T''(x) > 0$; therefore, the stationary point corresponds to a local minimum of the potential energy.

The local minimum is also a global minimum because at the boundaries of the domain, the total potential energy accepts equal values ($T(0) = T(d) = f(0) + f(d)$), and the first derivative of $T(x)$ given by equation 8.15 can turn into zero only for $x = d/2$, which means that there is a single stationary point in the interval [0,d].

$$T'(x) = f'(x) - f'(d - x) \tag{8.15}$$

The spring force is a conservative force. Its work depends on the length of the spring and does not depend on the trajectory by which the final length is attained or on the time. Consequently, the coordinates x,y and z of the end of a stretched spring define a conservative stationary field $\Phi(x, y, z)$. The work A in

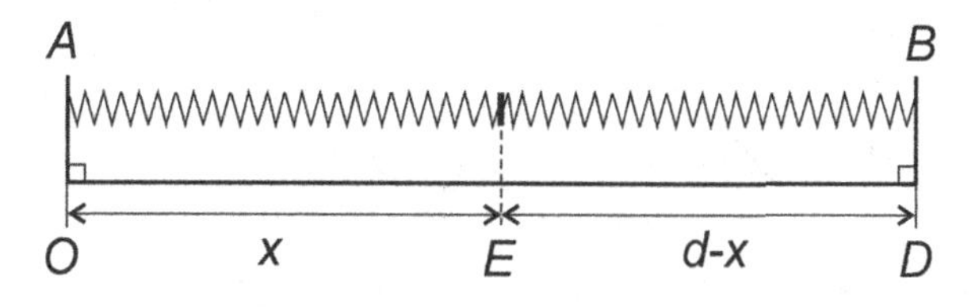

FIGURE 8.5 Physical interpretation of inequality (8.13) in terms of total potential energy of two non-linear springs.

a conservative stationary field $\Phi(x,y,z)$ depends only on the initial M_1 and final position M_2 in the field, but not on the trajectory (Figure 8.6).

The work A done in the stationary filed $\Phi(x,y,z)$ for moving an object from M_1 to M_2 is given by the line integral

$$A = \int_{M_1}^{M_2} (F_x dx + F_y dy + F_z dz) \tag{8.16}$$

where F_x, F_y, F_z are the components of the force $\mathbf{F}$ acting on the material point along the corresponding coordinate axes. The necessary and sufficient conditions for a stationary potential field are

$$F_x = \frac{\partial \Phi}{\partial x}, F_y = \frac{\partial \Phi}{\partial y}, F_z = \frac{\partial \Phi}{\partial z} \tag{8.17}$$

As a result, the elementary work done in a potential field is equal to the total differential of the potential field:

$$dA = \frac{\partial \Phi}{\partial x} dx + \frac{\partial \Phi}{\partial y} dy + \frac{\partial \Phi}{\partial z} dz = d\Phi \tag{8.18}$$

The total work done by the potential field for moving a material point from M_1 to M_2 is given by

$$A = \int_{M_1}^{M_2} d\Phi(x,y,z) = \Phi_2 - \Phi_1 \tag{8.19}$$

The work for moving the material point from M_1 to M_2 depends on the difference of values of the potential field at the initial and final points.

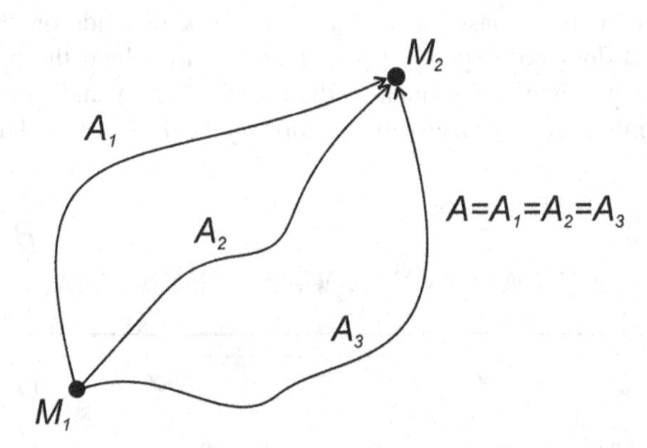

FIGURE 8.6 Conservative stationary field.

From equation (8.19), it also follows that the work around a closed path in the potential field is zero:

$$A_{0,0} = \int_{M_0}^{M_0} d\Phi(x,y,z) = \Phi_0 - \Phi_0 = 0 \tag{8.20}$$

The potential energy U is the work done by the conservative field for moving the material point from M_1 to M_2, taken with a negative sign. It expresses the storage of energy of the material point at a particular location in the conservative field.

Considering the definition of potential energy, the components of the force **F** acting on the material point become

$$F_x = -\frac{\partial U}{\partial x},\ F_y = -\frac{\partial U}{\partial y},\ F_z = -\frac{\partial U}{\partial z} \tag{8.21}$$

Going back to the system of springs in Figure 8.5, the forces F_1 and F_2 with which the springs act on the junction point E (Figure 8.5) are given by the derivatives of the potential energy of the spring:

$$F_1 = -\frac{df(x)}{dx};\ F_2 = -\frac{df(y)}{dy}$$

At the equilibrium point, $|F_1| = |F_2|$.

For a convex function, such as $f(x) = ke^{-x}$, where k is a positive constant, the inequality

$$ke^{-x} + ke^{-y} \geq 2ke^{-d/2} \tag{8.22}$$

where $x + y = d$ holds. The global minimum of the potential energy $T(x) = ke^{-x} + ke^{-(d-x)}$ is attained in the middle of the interval $[0,d]$ for $x=d/2$ and is equal to $2ke^{-d/2}$.

For potential energy given by the convex function $f(x) = k/(x+r)$, where k and r are positive constants, the inequality

$$k/(x+r) + k/(y+r) \geq 2k/(d/2+r) \tag{8.23}$$

where $x + y = d$ is also true. The global minimum of the potential energy $T(x) = k/(x+r) + k/(y+r)$ (the lower bound) is attained in the middle of the interval $[0,d]$ for $x=d/2$ and is equal to $2k/(d/2+r)$.

It must be pointed out that the lower bound of the total potential energy is not always attained in the middle of the interval $[0,d]$ for any function $f(x)$. For a concave function such as $f(x) = \sqrt{x+1}$, this is certainly not the case. The inequality

$$\sqrt{x+1} + \sqrt{y+1} \geq 2\sqrt{d/2+1} \tag{8.24}$$

where $x+y=d$, is not true. The minimum of the potential energy is attained at the ends of the interval [0,*d*] (for *x*=*0* or *x*=*d*) and is equal to $1+\sqrt{d+1}$. The correct inequality is

$$\sqrt{x+1}+\sqrt{y+1}\geq 1+\sqrt{d+1} \tag{8.25}$$

Despite that the spring forces $F_1=-\frac{d(\sqrt{x+1})}{dx}$ and $F_2=-\frac{d(\sqrt{y+1})}{dy}$ are equal in the middle, at point *E* (Figure 8.5), point *E* corresponds to a local maximum and the equilibrium at this point is unstable. A slight displacement from point *E* leads to a decrease of the total potential energy.

The developments presented in this book can be expanded significantly by interpreting known inequalities in the context of various specific domains: engineering, manufacturing, physics, economics and operational research. Particularly promising classes of inequalities suitable for meaningful interpretation are the inequalities based on sub-additive and super-additive functions. These can be readily interpreted if their variables and terms represent additive quantities.

References

Alsina C. and Nelsen R.B. (2010). *Charming Proofs: A Journey into Elegant Mathematics.* Washington, DC: The Mathematical Association of America.

Andriani P. and McKelvey B. (2007). Beyond Gaussian averages: Redirecting international business and management research toward extreme events and power laws. *Journal of International Business Studies*, 38, 1212–1230.

Ang A.H. and Tang W.H. (2007). *Probability Concepts in Engineering: Emphasis on Amplications in Civil and Environmental Engineering*, Hoboken, NJ: John Wiley & Sons.

Aven T. (2003). *Foundations of Risk Analysis*, Chichester: Wiley.

Aven T. (2017). Improving the foundation and practice of reliability engineering. *Proc IMechE Part O: Journal of Risk and Reliability*, 231(3), 295–305.

Bazovsky I. (1961). *Reliability Theory and Practice*, Englewood Cliffs, NJ: Prentice-Hall.

Bechenbach E. and Bellman R. (1961). *An Introduction to Inequalities*, New York: Random House, The L.W.Singer Company.

Ben-Haim Y. (2005). *Info-gap Decision Theory For Engineering Design. Or: Why 'Good' is Preferable to 'Best'*, in Engineering Design Reliability Handbook, Edited by E.Nikolaidis, D.M.Ghiocel and S.Singhal, Boca Raton, FL: CRC Press.

Berg V.D and Kesten H. (1985). Inequalities with applications to percolation and reliability. *Journal of Applied Probability*, 22 (3), 556–569.

Besenyei A. (2018). Picard's weighty proof of Chebyshev's sum inequality. *Mathematics Magazine*, 91(5), 366–371, DOI: 10.1080/0025570X.2018.1512814.

Budynas R.G. (1999). *Advanced strength and applied stress analysis*, 2nd ed., New York: McGraw-Hill.

Budynas R.G. and Nisbett J.K. (2015). *Shigley's Mechanical Engineering Design*, 10th ed. New York: McGraw-Hill.

Chandrasekaran R. and Tamir A. (1990). Algebraic optimization: The Fermat–Weber location problem. *Mathematical Programming*, 46(2), 219–224.

Cauchy A.-L. (1821). *Cours d'Analyse de l'École Royale Polytechnique, Première partie, Analyse Algébrique.* New York: Cambridge University Press.

Childs P.R.N. (2014). *Mechanical Design Engineering Handbook.* Amsterdam, The Netherlands: Elsevier.

Cloud M., Byron C., and Lebedev L.P. (1998). *Inequalities: With Applications to Engineering.* New York: Springer-Verlag.

Collins J.A. (2003). *Mechanical Design of Machine Elements and Machines.* New York: John Wiley & Sons.

Cover T. (1987). Pick the largest number, in *Open Problems in Communication and Computation*, edited by Cover T.M. and Gopinath B. New York: Springer-Verlag, 152–152.

DeGroot M.H. (1989). *Probability and Statistics*, 2nd ed. Reading, MA: Addison-Wesley Publishing Company.

DeVoe H. (2012). *Thermodynamics and Chemistry*, 2nd ed. Englewood Cliffs, NJ: Prentice-Hall.

Dhillon B.S. (2017). *Engineering Systems Reliability, Safety, and Maintenance.* Boca Raton, FL: CRC Press.

Dohmen K. (2006). Improved inclusion-exclusion identities and Bonferoni inequalities with reliability applications. *SIAM Journal on Discrete Mathematics*, 16(1), 156–171. DOI: 10.1137/S0895480101392630.

Easley D. and Kleinberg J. (2010). *Networks, Crowds, and Markets: Reasoning about a Highly Connected World*. New York: Cambridge University Press.

Ebeling C.E. (1997). *Reliability and Maintainability Engineering*. Boston, MA: McGraw-Hill.

Engel A. (1998). *Problem-Solving Strategies*. New York: Springer.

Fink A.M. (2000). An essay on the history of inequalities. *Journal of Mathematical Analysis and Applications*, 249, 118–134.

Floyd T.L. and Buchla D.L. (2014). *Electronic Fundamentals, Circuits, Devices and Applications*, 8th ed. London, UK: Pearson Education Limited.

French M. (1999). *Conceptual Design for Engineers*, 3rd ed. London: Springer-Verlag Ltd.

Gere J.M. and Timoshenko S.P. (1999). *Mechanics of Materials*. Cheltenham, UK: Stanley Thornes Publishers.

Gullo L.G. and Dixon J. (2018). *Design for Safety*. Chichester, UK: John Wiley & Sons.

Hardy G., Littlewood J.E., and Pólya G. (1999). *Inequalities*. Cambridge Mathematical Library, Cambridge, UK: Cambridge University Press.

Hearn E.J. (1985). *Mechanics of Materials*, 2nd ed., vol. 1. Oxford: Butterworth-Heinemann.

Henley E.J. and Kumamoto H. (1981). *Reliability Engineering and Risk Assessment*. Englewood Cliffs, NJ: Prentice-Hall.

Hill S.D., Spall J.C., and Maranzano C.J. (2013). Inequality-based reliability estimates for complex systems. *Naval Research Logistics*, 60(5), 367–374.

Horowitz P and Hill W. (2015). *The Art of Electronics*, 3rd ed. Cambridge, UK: Cambridge University Press.

Hoyland A. and Rausand M. (1994). *System Reliability Theory*. New York: John Wiley and Sons.

Kaplan S. and Garrick B.J. (1981). On the quantitative definition of risk. *Risk Analysis*, 1(1), 11–27.

Kazarinoff N.D. (1961). *Analytic Inequalities*. New York: Dover Publications.

Kundu C. and Ghosh A. (2017). Inequalities involving expectations of selected functions in reliability theory to characterize distributions. *Communications in Statistics – Theory and Methods*, 46(17), 8468–8478.

Levi M. (2009). *The Mathematical Mechanic*. Princeton, NJ: Princeton University Press.

Lewis E.E. (1996). *Introduction to Reliability Engineering 2nd ed.*, New York: John Wiley & Sons.

Livio M. (2009). *Is God a Mathematician?* New York: Simon and Shuster Paperbacks.

Makri F.S. and Psillakis Z.M. (1996). Bounds for reliability of k-within two-dimensional consecutive-r-out-of-n failure systems, *Microelectronics Reliability*, 36(3), 341–345.

Mannaerts S.H. (2014). Extensive quantities in thermodynamics. *European Journal of Physics*, 35(2014), 1–10.

Matthews C. (1998). *Case Studies in Engineering Design*. London: Arnold.

Miller I. and Miller M. (1999). *John E. Freund's Mathematical Statistics*, 6th ed. Upper Saddle River, NJ: Prentice Hall.

Modarres M., Kaminskiy M.P., and Krivtsov V. (2017). *Reliability Engineering and Risk Analysis, A Practical Guide*, 3rd ed. Boca Raton, FL: CRC Press.

Mott R.L, Vavrek E.M., and Wang J. (2018). *Machine Elements in Mechanical Design*, 6th ed. Upper Saddle River, NJ: Pearson Education.

Newman M.E.J. (2007). Power laws, Pareto distributions and Zipf's law. *Contemporary Physics*, 46(5), 323–351.

Norton R.L. (2006). *Machine Design, An Integrated Approach*, 3rd ed. Upper Saddle River, NJ: Pearson International.

O'Connor P.D.T. (2002). *Practical Reliability Engineering*, 4th ed. New York: John Wiley & Sons.

Pachpatte B.G. (2005). *Mathematical Inequalities*, North Holland Mathematical Library, vol. 67, Amsterdam, The Netherlands: Elsevier.

Pahl G., Beitz W., Feldhusen J., and Grote K.H. (2007). *Engineering Design*. Berlin, Germany: Springer.

Papadimitriou C.H. and Yannakakis M. (1982). *Combinatorial Optimization: Algorithms and Complexity*. Englewood Cliffs, NJ: Prentice-Hall.

Penrose R. (1989). *The Emperor's New Mind*. Oxford, UK: Oxford University Press.

Pop O. (2009). About Bergström's inequality. *Journal of Mathematical Inequalities*, 3(2): 237–242.

Ramakumar R. (1993). *Engineering Reliability, Fundamentals and Applications*. Upper Saddle River, NJ: Prentice Hall.

Rastegin A. (2012). Convexity inequalities for estimating generalized conditional entropies from below. *Kybernetika*, 48(2), 242–253.

Rosenbaum R.A. (1950). Sub-additive functions. *Duke: Mathematical Journal*, 17, 227–247.

Rozhdestvenskaya T.B. and Zhutovskii V.L. (1968). High-resistance standards. *Measurement Techniques*, 11, 308–313.

Samuel A. and Weir J. (1999). *Introduction to Engineering Design: Modelling, Synthesis and Problem Solving Strategies*. London, UK: Elsevier.

Sedrakyan H. and Sedrakyan N. (2010). *Algebraic Inequalities*. Cham, Switzerland: Springer.

Steele J.M. (2004). *The Cauchy-Schwarz Master Class: An Introduction to the Art of Mathematical Inequalities*. New York: Cambridge University Press.

Tegmark M. (2014). *Our Mathematical Universe*. London, UK: Penguin books.

Thompson G. (1999). *Improving Maintainability and Reliability through Design*. London, UK: Professional Engineering Publishing.

Tipler P.A. and Mosca G. (2008). *Physics for Scientists and Engineers: With Modern Physics*. New York: W.H. Freeman and Company.

Todinov M.T. (2002a). Statistics of defects in one-dimensional components. *Computational Materials Science*, 24, 430–442.

Todinov M.T. (2002b). Distribution mixtures from sampling of inhomogeneous microstructures: Variance and probability bounds of the properties. *Nuclear Engineering and Design*, 214, 195–204.

Todinov M.T. (2003). Modelling consequences from failure and material properties by distribution mixtures. *Nuclear Engineering and Design*, 224, 233–244.

Todinov M.T. (2006a). Equations and a fast algorithm for determining the probability of failure initiated by flaws. *International Journal of Solids and Structures*, 43, 5182–5195.

Todinov M.T. (2006b). Reliability analysis of complex systems based on the losses from failures. *International Journal of Reliability, Quality and Safety Engineering*, 13(2), 1–22.

Todinov M.T. (2007). *Risk-Based Reliability Analysis and Generic Methods for Risk Reduction*. Amsterdam, The Netherlands: Elsevier.

Todinov M.T. (2013). New models for optimal reduction of technical risks. *Engineering Optimization*, 45(6), 719–743.

Todinov M.T. (2016). *Reliability and Risk Models: Setting Reliability Requirements*, 2nd ed. Chichester, UK: John Wiley & Sons.

Todinov M.T. (2017). Reliability and risk controlled by the simultaneous presence of random events on a time interval. *ASCE-ASME Journal of Risk and Uncertainty in Engineering Systems, Part B: Mechanical Engineering*, 4(2), 021003. DOI: 10.1115/1.4037519.

Todinov M.T. (2019a). Domain-independent approach to risk reduction. *Journal of Risk Research*, 23, 796–810. DOI: 10.1080/13669877.2019.1628093.

Todinov M.T. (2019b). *Methods for Reliability Improvement and Risk Reduction*. Hoboken, NJ: John Wiley & Sons.

Todinov M.T. (2019c). Improving reliability and reducing risk by using inequalities. *Safety and Reliability*, 38(4), 222–245. DOI: 10.1080/09617353.2019.1664129.

Todinov M.T. (2019d). Reliability improvement and risk reduction by inequalities and segmentation. *Proc IMechE Part O: Journal of Risk and Reliability*. DOI: 10.1177/1748006X19869516.

Todinov M.T. (2020a). *Risk and Uncertainty Reduction by Using Algebraic Inequalities*. Boca Raton, FL: CRC Press.

Todinov M.T. (2020b). On two fundamental approaches for reliability improvement and risk reduction by using algebraic inequalities. *Quality and Reliability Engineering International*. DOI: 10.1002/qre.2766.

Todinov M.T. (2020c).Using algebraic inequalities to reduce uncertainty and risk. *ASCE-ASME Journal of Risk and Uncertainty in Engineering Systems, Part B: Mechanical Engineering*, 6(4). DOI: 10.1115/1.4048403.

Todinov M.T. (2020d). Reducing uncertainty and obtaining superior performance by segmentation based on algebraic inequalities, *International Journal of Reliability and Safety*, 14(2/3), 103–115.

Todinov M.T. (2020e). Generation of new knowledge and optimisation of systems and processes through meaningful interpretation of sub-additive functions. *International Journal of Mathematical Modelling and Numerical Optimisation*, to appear.

Todinov M.T. (2021). Meaningful interpretation of algebraic inequalities to achieve uncertainty and risk reduction. *Proc IMechE Part O: Journal of Risk and Reliability*, published online, doi/full/10.1177/1748006X211036573.

Vedral V. (2010). *Decoding Reality*, Oxford, UK: Oxford University Press.

Vose D., (2000). *Risk Analysis, A Quantitative Guide*, 2nd ed. New York: John Wiley & Sons.

Wesolowsky G. (1993). The Weber problem: History and perspective. *Location Science*, 1, 5–23.

Wigner E., (1960). The unreasonable effectiveness of mathematics in the natural sciences, *Communications in Pure and Applied Mathematics*, 13(1).

Winkler R.L. (1996). Uncertainty in probabilistic risk assessment. *Reliability Engineering and System Safety*, 85, 127–132.

Weiszfeld E. (1937). Sur le point pour lequel la somme des distances de *n* points donnes est minimum. *Tohoku Mathematical Journal*, 43, 355–386.

Wolfson R. (2016). *Essential University Physics*, 3rd ed. Upper Saddle River, NJ: Pearson.

Xie M. and Lai C.D. (1998). On reliability bounds via conditional inequalities. *Journal of Applied Probability*, 35(1), 104–114.

Index

Printed in the United States
by Baker & Taylor Publisher Services